Mohammad Samsamshariat

Product development of earthquake-safe houses and schools

Mohammad Samsamshariat

Product development of earthquake-safe houses and schools

With focus on developing countries

Südwestdeutscher Verlag für Hochschulschriften

Impressum/Imprint (nur für Deutschland/ only for Germany)
Bibliografische Information der Deutschen Nationalbibliothek: Die Deutsche Nationalbibliothek verzeichnet diese Publikation in der Deutschen Nationalbibliografie; detaillierte bibliografische Daten sind im Internet über http://dnb.d-nb.de abrufbar.

Alle in diesem Buch genannten Marken und Produktnamen unterliegen warenzeichen-, marken- oder patentrechtlichem Schutz bzw. sind Warenzeichen oder eingetragene Warenzeichen der jeweiligen Inhaber. Die Wiedergabe von Marken, Produktnamen, Gebrauchsnamen, Handelsnamen, Warenbezeichnungen u.s.w. in diesem Werk berechtigt auch ohne besondere Kennzeichnung nicht zu der Annahme, dass solche Namen im Sinne der Warenzeichen- und Markenschutzgesetzgebung als frei zu betrachten wären und daher von jedermann benutzt werden dürften.

Verlag: Südwestdeutscher Verlag für Hochschulschriften GmbH & Co. KG
Dudweiler Landstr. 99, 66123 Saarbrücken, Deutschland
Telefon +49 681 37 20 271-1, Telefax +49 681 37 20 271-0
Email: info@svh-verlag.de
Zugl.: Wuppertal, Bergische Universität Wuppertal, Diss. , 2009

Herstellung in Deutschland:
Schaltungsdienst Lange o.H.G., Berlin
Books on Demand GmbH, Norderstedt
Reha GmbH, Saarbrücken
Amazon Distribution GmbH, Leipzig
ISBN: 978-3-8381-1837-6

Imprint (only for USA, GB)
Bibliographic information published by the Deutsche Nationalbibliothek: The Deutsche Nationalbibliothek lists this publication in the Deutsche Nationalbibliografie; detailed bibliographic data are available in the Internet at http://dnb.d-nb.de.

Any brand names and product names mentioned in this book are subject to trademark, brand or patent protection and are trademarks or registered trademarks of their respective holders. The use of brand names, product names, common names, trade names, product descriptions etc. even without a particular marking in this works is in no way to be construed to mean that such names may be regarded as unrestricted in respect of trademark and brand protection legislation and could thus be used by anyone.

Publisher: Südwestdeutscher Verlag für Hochschulschriften GmbH & Co. KG
Dudweiler Landstr. 99, 66123 Saarbrücken, Germany
Phone +49 681 37 20 271-1, Fax +49 681 37 20 271-0
Email: info@svh-verlag.de

Printed in the U.S.A.
Printed in the U.K. by (see last page)
ISBN: 978-3-8381-1837-6

Copyright © 2011 by the author and Südwestdeutscher Verlag für Hochschulschriften GmbH & Co. KG and licensors
All rights reserved. Saarbrücken 2011

This work is dedicated to the people, who lost their life to an earthquake disaster, and to their surviving dependants.

Acknowledgements

This dissertation is being written during my PhD at Bergische University Wuppertal with the financial support of German Academic Exchange Service (DAAD). I would like to deeply appreciate the well organized scholarship as well as the whole supports offered by DAAD during this period.

This thesis is part of a bigger project aiming at affordable earthquake-safe houses for developing countries. I like to express my special thanks to my supervisor, Univ.-Prof. Dr.-Ing. Georg Pegels, who made his support available in many ways during this project and my work. Without his comments and suggestions in edition of the dissertation, this thesis would not have been possible.

Special and sincere thanks are given to all members of the board of examiners, Univ.-Prof. Dr.-Ing. Harte and Univ.-Prof. Dr.-Ing. Schlenkhoff from Bergische University Wuppertal and Dr. Parvizian from Isfahan University of Technology for attending the exam and for their comments.

I would like to express my special gratitude to Mr. Dipl.-Ing. Torsten Weckmann, managing director of the engineering company ICW GmbH, for sharing his ideas and experiences during several discussions we had and for giving encouragement to complete the work.

Many other friends and colleagues helped accomplishing this work with their useful recommendations. Special thanks to my fellow and colleague, Dr.-Ing. Hamid Reza Azadnia for his constructive comments on the whole work and for his experience in CAD/CAM and steel detail engineering. This work would not have been successful without his assistance. I also like to thank Mrs. Shabnam Kabiri, MSc, for her helps during my PhD.

I am grateful to Ms. Nahid Nasserian, MSc, for all her creative architectural ideas and for contribution to the work with her valuable suggestions.

Last but not least, I wish to express my deepest gratitude to my beloved parents for the whole support they have always given to me. Their kind advises is been my highest motivation during my thesis and in whole my life.

Abstract

Millions of people worldwide are in urgent need of affordable houses. The majority of these populations are either first-time home owners or those who have lost their dwellings as a result of a natural hazard, like an earthquake. According to the statistics, these people are mainly living in developing countries with a relatively high level of seismic risk and low to medium incomes.

In addition to the residential buildings, schools are one of the most demanded infrastructures with a high level of importance in sense of seismic design; their damage endangers a high number of occupants, and interrupts their function as urgent shelters in afflicted areas.

While traditional practices are unable to cope with currently high demands, large number of fatalities and wide spread damages during last earthquakes show that houses in many regions of the world are seismically vulnerable. The high level of vulnerability is mainly caused due to inappropriate construction techniques and reveals that houses and schools have to be designed, detailed and constructed with an adequate level of earthquake-resistance and an affordable price.

This work aims at developing a simple and low-cost solution, inspired by good performance of traditional low-rise half-timber structures. Therefore, the main parameters of this good performance is realized and adopted for the recommended solution, called systematically braced frames. Due to the multidisciplinary nature of the problem, a framework is developed by innovative adoption of Product Development into civil engineering, which prepares the basis for contribution of several required disciplines, even in future. The solution is then developed during four phases of Product Development, namely:

- Task clarification,
- Conceptual design of main components,
- Structural design of the steel structure, and
- Detail design of a sample school building in Iran to investigate the significance and applicability of the solution.

To assure a high quality, an off-site prefabrication of main safety-relevant structural elements under strict supervision is supported here, while the rest of the structure can be accomplished by local labor, to generate local job opportunities, using regional available materials, to provide cultural acceptance and to minimize costs.

As required for the seismic study, horizontal stiffness and ductility are investigated analytically during the work.

Contents

Title	Page

1. **Introduction** 7
   - 1.1. Motivation and target of the work 7
   - 1.2. Problem definition 9
   - 1.3. Methodology and innovation 10
   - 1.4. Limits of the study 12
   - 1.5. Organization of the thesis 12
2. **Basics of product development and task clarification** 15
   - 2.1. Basics of product development 15
   - 2.2. Modularization of products 23
   - 2.3. Construction methods 25
3. **Market survey** 27
   - 3.1. Housing demand 27
   - 3.2. Demand on schools 30
   - 3.3 Seismic risk 31
   - 3.4 Major construction types in developing countries 37
4. **Last earthquakes: costs paid, lessons learnt** 47
   - 4.1. Architectural related aspects 47
   - 4.2. Structural engineering aspects 53
   - 4.3. Constructional aspects 62
5. **Conceptual design of earthquake-safe houses and schools** 65
   - 5.1. Structural concept 65
   - 5.2. Structural safety-relevant components 68
   - 5.3. Cost analysis and calculation 85
   - 5.4. Non-structural components 89
6. **Structural design of earthquake-safe houses and schools** 93
   - 6.1. Structural features of the main components 93

	6.2. Other important aspects...	102
7.	**Case study: A sample school in Iran**	**113**
	7.1. Building description..	113
	7.2. Modeling and analysis...	115
	7.3. Cost calculation..	120
	7.4. Highlights..	120
8.	**Concluding remarks and future works**..	**123**
	References...	**125**

1. Introduction

1.1. Motivation and target

Earthquakes cause high death tolls and wide spread damage in the world every year, (figure 1-1). In addition to the hundreds of thousands of fatalities in the last century, millions have been left homeless as a result of this natural hazard, many of whom can not afford a new life under current economic circumstances. These people are deprived of their fundamental rights. Article 25 of the United Nations Universal Declaration of Human Rights (1948) states that "Everyone has the right to a standard of living adequate for the health and well-being of himself and of his family, including food, clothing, housing and medical care and necessary social services...."

Figure 1-1: Collapse of school buildings in Wenchuan, China during the M 7.9 earthquake of May 12, 2008 [13]

In addition to the high level of seismic hazard in developing countries, structural and social vulnerability to hazard is also high. Besides, there is an increasing tendency in developing countries to migrate to the mega cities, even to the earthquake prone ones, to seek for appropriate jobs. The United Nations estimates that by 2015, 23 cities will have populations exceeding 10 million, and of those, all but 4 will be in less developed countries. Of the top ten urban agglomerations projected for 2015, eight are cities with a known moderate to high seismic risk, including Tokyo, Mumbai, Dhaka, Karachi, Mexico City, New York, Jakarta, and Calcutta. A major earthquake in one of these cities, particularly in a city with vulnerable buildings and fragile infrastructures, could cause catastrophic devastation and millions of deaths. A fitting example of this case is the metropolitan area of Tehran. Currently, 13 million inhabitants of Tehran live in houses of which more than half would not be habitable after an earthquake of 6 on the Richter magnitude scale. And Iran, with 126,000 fatalities since 1900 and 17 earthquakes with more than 1000 deaths each during this period is one of the most vulnerable countries in terms of earthquake hazard [1].

To reduce poverty in developing countries despite the large population growth, a huge number of jobs and houses have to be created every year. Both problems, inadequate housing and insufficient jobs, have become so important that the United Nations has called for their urgent solution as a Millennium Development Goal, set to be achieved by 2015 by all members. While earthquakes happen mostly in developing countries with high population densities and low incomes, mass production of low-cost housing units looks a promising solution.

Appropriate and systematic development of the housing industry provides not only houses for millions in need, but will also create an enormous number of jobs and lead to a major development. One of the best examples happened historically in Germany after the Second World War. This issue has recently grabbed the attention and interest of some politicians in emerging markets of developing countries. Dr Mahathir bin Mohamad, the former Prime Minister of Malaysia mentions in [2], that knowledge-based technologies give developing countries excellent opportunities to create job places. Therefore, entering into the information age, which is already started, is the most radical economic change since the industrial revolution. He believes that businesses can create enough employment by using this opening in different phases:

- Converting information into application;
- Finding and using potentials and existing capacities;
- Combining technology with public opinion;
- Developing simple ideas and continuous improvement of existing products;
- Training skilled workers and engineers.

As he explains, being a developing country apparently does not hinder the development of knowledge-based businesses. These businesses span the whole spectrum and any of thousands of industries involved in a big business can be chosen in order to propel knowledge-based commerce in a developing country. To determine what kind of business and what niche is the correct one for a country, the specific assets of that country should be identified. Knowledge of these assets and the capacity of modern technology will suggest the type of industry that country can go into. Businesses that need very big investments are obviously unsuitable for developing countries, while some others costing very little can bring high returns. For instance, transfer of knowledge and technology in the building industry seems very suitable for developing countries. Because of the high and continuous demand on housing, these countries can fulfil their own requirements while creating job opportunities and entering to the world market by cooperation.

1.2. Problem definition

The gap between developed and developing countries is widening in terms of seismic vulnerability and buildings are constructed still with high vulnerability in many earthquake prone areas. Consequently, four of every five deaths caused by earthquakes in the twentieth century occurred in developing countries. Of people living in earthquake threatened cities in 1950, two out of every three were in developing countries; in 2000, nine out of ten were in developing countries [3].

Despite the investments and the progress of research in earthquake engineering worldwide, experience in recent earthquakes, particularly in developing countries demonstrates that, so far, earthquake engineering has been unable to get rid of the unexpectedly high number of deaths. Editors of the World Housing Encyclopaedia of the Earthquake Engineering Research Institute and the International Association of Earthquake Engineering have recently pointed out the most prominent challenge of earthquake engineering [4]. In their opinion, effort is called for to reduce seismic vulnerability of structures in developing countries. They consider that one of the ways to accomplish this goal is to encourage researchers worldwide to adopt and apply knowledge into practice to solve real problems. For this purpose, low-rise buildings, in which most fatalities and damage have occurred in recent earthquakes, should become the focus of research.

On the other hand, the seismic vulnerability of new and existing buildings must be reduced in developing countries. The currently high and growing vulnerability in developing countries is not mainly due to a lack of knowledge or technical capabilities, but to the lack of strong policy implementation, knowledge application and cooperation, as well as limited economic capabilities and awareness [5]. In other words, the roots of the problem are in the socio-political rather than the engineering frame, and here the ground is ripe for corruption. For instance wrong construction practices in the lack of successful supervision, which increases the damage and death risk [6]. In addition, the traditional housing construction practices, which are still in use in developing countries, cannot fulfil the current high demand in this sector, neither quantitatively nor qualitatively. The upshot is that, because governments and authorities have not been able to solve the problem efficiently, dealing with the problem at all is becoming a taboo in many countries.

The key role of civil engineering in public welfare is evident in the fact that it can revolutionize this process and simultaneously create new job opportunities for a growing population. Overcoming both challenges with one systematic approach is a key turning point for all developing countries affected by these problems. However, meeting the demand for hundreds of thousands of houses per year with high quality and individuality is only possible if structural elements of the houses are manufactured in the factory in an industrial way and under strict supervision. Therefore, automated production of safety-relevant structural components of houses with high quality is indispensable. Besides increasing the quality and speed, prefabrication also reduces the final price effectively and makes houses affordable, even for families with low- and moderate-incomes in rural areas and small cities.

While today prefabricated housing is becoming a practical means of increasing the quality level of the market, it still has not taken off in developing countries.

Selection and appropriate use of local construction materials are other important aspects affecting both the affordability and acceptance of residential buildings. There is no doubt that among the currently available materials, steel is unique in the sense of environmental sustainability. It can be easily produced in different shapes and can also be formed mechanically after casting. Besides, its high ductility provides the structure with a high performance in earthquake conditions. Nowadays, steel is increasingly available in different regions, including developing countries and steel factories are easily able to provide the required steel for the housing industry. Using steel instead of its main alternative, reinforced concrete, will solve the existing problems, which appear to be due to the relatively high stiffness and low ductility of concrete structures.

After production, prefabricated elements can be transported to the site for construction. Local labour is able to fit the infill walls and finishing coatings on site with existing traditional techniques. This way, the final costs will be reduced, and the building will be more acceptable, since its appearance is adapted to local taste and culture.

1.3. Methodology and innovation

Since the problem involves different aspects and these aspects have to be solved simultaneously, a multidisciplinary and interdisciplinary solution has to be developed here. An appropriate frame helps to prepare the ground for the contribution of a variety of disciplines to solve a real-life problem. In the present work, the solution is worked in the frame of product development, considering all factors of influence. Although this is a well-known method of manufacturing products to fulfil market needs, its introduction to the housing and school construction industry is new. Product development is used classically in mechanical engineering, where a specific product is manufactured in a large mass. Hence, the fundamentals of the design work will be built to best advantage.

In the classical approach, residential and public buildings are calculated and constructed separately and alone for each case, and the whole process of design is repeated for each building. Obviously, this approach has taken a large amount of investment and effort over the years, but the explanation has been to end up with distinctive and unique products. Recently, the market leading companies have changed the playing rules by utilization of prefabrication technology into the building industry, still having distinctive and unique products. While the components are from the same family, the improving factors over the time are the quality and efficiency of production and erection, the final product looks individual and different from the others. As it is evident in figure 1-2, some enterprises like Goldbeck in Germany have been even so successful in these terms to broaden their business to other countries, although labour costs have been lower there [17]. Interestingly, implementation of many goals of product development, for instance, meeting customers' needs, design for quality and design for minimum costs as well as modularization of components are an evidence of a successful product development in this enterprise.

1. Introduction

In this work, adoption of product development for design of earthquake-safe houses and schools is a new and innovative approach. The main structural problems of these types of buildings are found out and listed to be solved. Afterwards, the question is addressed by implementation of systematically braced frames, which are the modification of traditional half-timber structures in Europe. Moreover, their main safety-relevant structural elements including floors and roofs as well as the load bearing wall panels, are designed. As it is the case in every new product development practise, the solution includes a new combination of working functions, which have to be explained by scientific rules. This has been done here to investigate a very important concept of the seismic design, namely ductility. In order to set product development goals, design criteria that the buildings and houses have to be judged by are listed and explained. Finally, different details of the buildings are worked through by CAD/CAM. It should be mentioned here that the advantage of the developed frame within this dissertation is that the effort required to implement some changes in the design is minimized, because it has to pass the existing steps. In other words, the systematic approach established under product development, makes it possible to get rid of deficiencies which happen often in the buildings designed and constructed with the classical approach and to improve continuously the product, in an efficient way.

Figure 1-2: A Gobacar® multi-storey car park and the production line in Treuen, Germany, Company turnover on construction abroad [17]

1.4. Limits of the study

The main focus of this work is on low-rise buildings, i.e. family houses and schools up to three floors without lift, which are currently the most predominant building type accommodating more than 80% of the world's population. Consideration of schools together with houses is important, because schools are one of the most demanded infrastructures and every neighbourhood needs to posses at least one school. In addition, the public profile of schools will prepare the market more for residential buildings, in the sense of acceptability. Considering their publicity, schools can be an excellent teaching example how to build safe houses. However, schools make different demands than houses due to their different usage and structure, and are therefore considered in a different category in this work.

Arguments and solutions have been intentionally kept as general as possible. Nevertheless, in several cases there has been a need for a specific example. In such cases the required data are used from Iran, the home country of the author. To implement the method for different cases, the validity and similarity of background conditions must be checked. Also building codes and standards used in this work are European Codes, prepared and accepted by member states of Europe. The advantage of these codes is that their application in a third country is accounted for systematically, simply by changing the relevant coefficients. In other words, where a coefficient can vary from one region to another, this fact is explicitly noted. Therefore, because the general method is the same everywhere, the international cooperation is facilitated.

The present thesis will address product development of prefabricated seismic-resistant houses and schools, considering local acceptability. The houses in question are designed for rural areas and small to medium-sized towns rather than mega cities. Hence, they do not need to have more than three storeys and lift.

At the end of the study the following hypothesis will be investigated:

- Is a prefabricated steel structure appropriate for developing countries with a high level of seismic hazard, to be used for houses and schools?
- Could high technology production and construction methods overcome the existing barriers of the housing market in developing countries?
- Are all factors of influence considered to guarantee safe results?

1.5. Organization of the thesis

This dissertation is divided into eight chapters. The present chapter defines the existing problem and the aspects to be considered. Also the methodology has been presented briefly. In chapter two basics of product development are explained and the required tasks are listed for different components of houses and schools. In the third chapter, an overview on the level of seismic hazard and its consequences on houses and schools in developing countries are presented. Since it is very important to know the requirements from the early stages, the economical situation is also evaluated in this chapter. A list of the most damage-causing reasons investigated in recent earthquakes in buildings is presented in chapter four.

The process of product development of houses and schools is the focus of the next three chapters. Chapter five addresses the conceptual design of safety-relevant components of houses and schools. Structural design of the products is the subject of sixth chapter, where the main required theories are developed and discussed. To show the applicability of the thesis, as a case study, a school in Iran is designed in chapter seven.

Finally, the conclusions obtained during the work are mentioned in chapter eight. This chapter contains also recommendations for further works.

2. Basics of product development and task clarification

2.1. Basics of product development

Product Development is the process of preparing and bringing a product or a service to the market with the right price and right performance at the right time. This involves all rational ways necessary to make complex processes comprehensible and transparent. The systematic approach of product development aims at reducing the design iteration loops and to make them as effective and as efficient as possible to offer products with better quality and better appearance as any other competitor. Figure 2-1 shows the iterative path from the existing task or problem to a desired solution in the whole product development process.

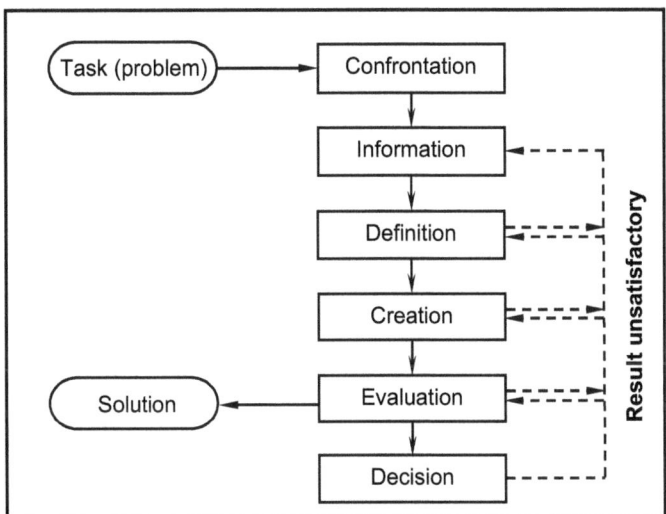

Figure 2-1: General problem solving process [35]

To be successful in the present competitive world, new products have to be designed and developed with special care to the following factors [35]:

- All functions expected from the new product must be known and all necessary activities needed to reach these functions must be realistically evaluated;
- A correct timing and scheduling of all these activities must be planned;
- The project and product costs must be continuously monitored and optimized.

Although implementation of these goals depends very much to the type of the product, developers have established some common steps, called also phases, to reach their target in the most effective way. The four main phases of product development are as below [35]:

1. Planning and task clarification, market analysis and customers' demands
2. Conceptual design, innovative ideas, brainstorming
3. Embody or structural design
4. Detail design

These phases include in turn some steps that are more or less similar. The common steps below describe the work flow inside and between these phases.

2.1.1. Clarification of the requirements

The first strategic step is to understand for which markets we want to develop products. In this step, the main task is defined by a market analysis and by feedbacks from customers. Based on the main task, all requirements that the product has to fulfill are listed. The importance of this stage is to go through the ocean of opportunities and find the gaps in the market. The good ideas are then reviewed and screened by experts in order to present the concept brief and pass to the next stage. This list is used as the main document in the design process later on. However, this list will be updated continuously as the project makes progress. One of the possible ways to list the project requirements is to prepare a product requirement matrix. In this matrix customer requirements are listed versus the product requirements. This way, the team can have a clear understanding of the tasks to be done. For the main purpose of this work, earthquake-safe houses and schools, a set of requirements was established at the beginning. The main requirements list is prepared not only for the whole product, but also for its parts. In the case of houses and schools, these parts mean the main load bearing components, which are walls as well as floors and roofs. In [84], requirement tables are presented for wall and floor/roof assemblies according to the European standards. These are presented in table 2-1 and 2-2, respectively, with minor modifications. One of the modifications was to include seismic-resistance criteria in the tables.

Earthquake-safety

According to the arguments presented in chapter 3, target markets are largely in earthquake-prone regions of the world. Consequently, structures must be definitely earthquake-resistant. It is also mentioned in chapter 4, that earthquake-safety cannot be provided only by a structural designer. Earthquake-safety of an improper planned building with many asymmetries, or a building with a proper plan and structural design, but with a poor construction, becomes very expensive, if possible at all.

Currently, a variety of terms are used describing the design of buildings for seismic actions. Earthquake-proof design seems to be the first description in codes. However, this word can be found today less in new versions of codes. On the other hand earthquake-resistant or seismic-resistant design is the most frequently used term and can also be found worldwide in almost all codes considering this issue. Recently, earthquake-resilience is used by some initiations, which take also reconstruction and the social issues of earthquake into consideration. In the present work, the fact that dealing with earthquake hazard is a public duty and lies not only in the hands of a structural engineer is supported. Therefore, it is recommended to use the more general term, earthquake-safety which seems to have the appropriate concept for this work. Other definitions are also used where the specific meaning is of importance.

Light weight components

The components of a building have to be light enough for two reasons. The first reason is that light weight components ease and speed up the construction and assembling process. For villages with difficult access, light weight components can be transported and erected without any need to big trucks or cranes. The second reason is that light weight components lead to a light weight building, which is decisively important for earthquake-resistance, since the seismic action is a function of the weight multiplied by ground acceleration.

Visibility and provability

Since there is a lack of supervision in many rural and urban regions to control the correctness of the construction activities, the technical solution must eliminate the chance of corruption to occur. In [6] the high effect of corruption on increasing seismic risk is discussed and investigated. As experience has shown in last earthquakes, whenever quality and quantity of the material is not provable, corruption may occur. For example, it is very difficult to check the existence of enough reinforcing steel in the concrete. Lack of enough reinforcing steel in concrete structures has been frequently a reason of collapse. In contrary, the possibility to control the quality and the quantity of structural components at any time hinders the corruption chance to a great deal. Visibility seems to be the easiest way of controllability. However, composing of structural elements in order to make them visible will need engineering competence to prevent conflict with other requirements. For example, exposed steel components in buildings can create unsuitable thermal conditions of the building, and structural safety against fire must be considered.

Ability to accomplish non-safety-relevant works

Another aspect of cost-effectiveness and cultural acceptance is to take the advantage of local material. However, the type and also quality of material are very different from a place to another. Therefore, the structure must be earthquake-resistant by itself and must take the least effect from infill materials. This issue seems to be satisfied if infill materials have low strength and stiffness properties but high damping quality. Nevertheless, for earthquake-resistant buildings, it can happen that the structure is planned and designed for a soft behavior, which is advantageous for earthquake-resistance, but the building is made stiff by infill materials and gets therefore more seismic forces. Therefore, this aspect should be investigated and different infill materials should be considered in product development [41].

Affordability

As explained in chapter 3, low- to medium-income is one of the specific properties of the customers. Thus, the total cost of products must be affordable for them. Not only buying price, but also extra costs including additional charges and maintenance must be kept low. Furthermore, as shown by successful enterprises like Goldbeck in Germany, the enterprise should help costumers, to finance their object. However, financing have different means and methods depending on the type of the object. This issue is discussed in more detail in chapter 6.

Property or restraint	Requirement in European Standards
Imposed loads	Load effects have to be considered in design in accordance with EN 1991-1.
Dead loads	The maximum deflection of horizontal framing members must not exceed the minimum of L/500 and 3 mm.
Resistance to wind Load	Wall panels must have enough rigidity against positive, negative and cyclic effects of the wind load.
Resistance against seismic loads	Wall panels must have enough rigidity to limit the drift in each storey to the damage limitation requirement of EN 1998-1. Each wall assembly must be able to be pushed to is target displacement, defined in EN 1998-1, in is own plane, without collapse or instability and with acceptable seismic performance. Each wall assembly must be able to support the target displacement the storey belongs to, perpendicular to is own plane, without collapse or instability. Each wall assembly must have an acceptable ductility and energy dissipation capacity.
Thermal transmittance	Taking account of the varying external thermal conditions including extreme temperatures based on the systematic of EN ISO 10077-1. Walls should accommodate layers of thermal and acoustic insulation materials without high increase in thickness.
Air permeability	The performance of the product shall be within the limit of the class selected from PrEN 12152.
Water tightness	The performance of the product shall be within the limit of the class selected from PrEN 12154.
Water vapor permeability	To be controlled by the appropriate use of water vapor barriers.
Fire resistance	Ability to prevent the flow of heat and smoke according to DIN 4102. Taking account of structural fire design rules of EN 1993-1-2.
Acoustic insulation	The wall shall achieve the levels of acoustic insulation specified in EN (WI) 33216. The wall has to insulate the room from outer noises.
Thermal movement	The wall should accommodate thermal movements without causing extra actions to the structure.

Table 2-1:   Product requirements of walls [84]

2. Basics of product development and task clarification

Property or restraint	Requirement in European Standards
Building movement	The wall panel should damp movements caused by earthquake or wind. In case of using seismic base isolators, the wall should absorb the shock from changing movement direction according to EN 1998-1.
Resistance against impact	Where specifically required, the wall should comply with the requirements of EN (WI) 33218.
Durability	The wall panel including its connections should retain its structural integrity in design life of the building. The wall elements must be protected against corrosion e.g. by galvanization; special considerations should be taken into account according to EN ISO 1461, EN ISO 14713 and EN ISO 10684.
Dangerous substances	Materials should not release dangerous substances in excess of the maximum permitted levels specified in European material standards and other relevant material regulations. The main materials must be recyclable and the production and recycling process should emit the minimum amount of CO_2.
Transport	Dimension of elements including package and safety devices must not exceed the maximum deliverable limits defined by national transportation regulations.
Assembly	Wall elements should be light enough to be carried by few persons at the site, without need of crane. Elements should have no cutting edge, which makes carrying difficult.
Affordability	The construction and infill of walls must be possible with local materials and unskilled labor.
Controllability	Existence and quality of decisive structural elements must be verifiable by owners by eyes or simple tools to prevent corruption.
Acceptability and attractiveness	Elements must have the ability of being covered by different facade elements with optional materials. Wall assemblies should have the flexibility of being combined with each other with special care to local culture and architecture.

Table 2-1: Product requirements of walls [84] (Continued)

2.1.2. Establishing working sub-structures

Here, the overall function, which must be fulfilled by the product, is divided into sub-structures of essential and additional task-specific modules. It is important that different sub-structures are physically and logically compatible. In figure 2-2 an example is presented from [40], in which different variants of sub-structures and their combinational possibilities are shown for a simple building.

After studying market expectations, variants in low demand and/or with high overall costs will be eliminated or added as customer-specific functions. This case happens for instance in choosing the appropriate variants for walls. As a result of new progresses in the field of modern constructional material, a variety of new variants like hollow material and foam wall elements are recently available for residential buildings with an ultra light weight. Nevertheless, most of the house owners prefer massive elements even if they have less or no advantage compared to modern elements. Therefore, the option of hollow or foam walls can be eliminated from the variants. This helps achieving a more cost-effective and attractive product.

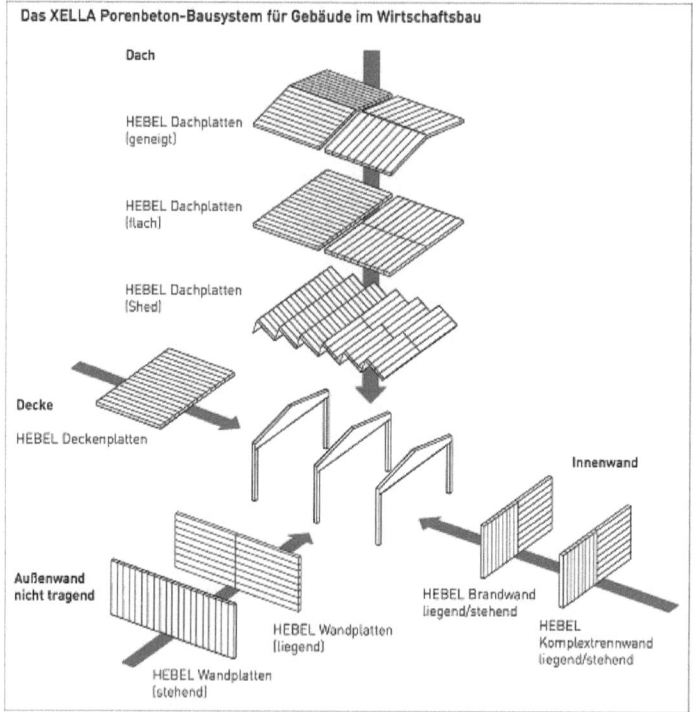

Figure 2-2: Schematic presentation of sub-structural variants for a simple building [40]

As mentioned before, it is generally preferred to implement the overall function by a minimum number of assemblies. It is also recommended to implement the main functions by essential modules and other functions by additional, special and adaptive functions. Very rarely demanded functions can be implemented as non-modules. It is more cost-efficient, if several functions are implemented by one module, wherever possible [35]. This is the case for instance in design of infill walls, because they serve at the same time for thermal and acoustic isolators. Properties of different infill walls for earthquake-safe schools and houses are explained in detail in [41].

2. Basics of product development and task clarification

Property or restraint	Requirement in European Standards
Imposed loads	Load effects must comply with EN 1991-1.
Dead loads	Floor/ roof elements should be manufactured to have a minimum weight; this is of main advantage for the economy of the final product.
Resistance to wind Load	The roof connections should have enough resistance against wind suction.
Resistance against seismic loads	Roof elements and connections must form an integrated floor/ roof element. The roof must guarantee the diaphragm behavior according to EN 1998-1, which means the distribution of horizontal loads to the wall elements are proportional to their stiffness. Connection of roof to the walls must have rotation capacity to support the required inter-story drift.
Vibrations	The Eigen frequency of the floor should be above 10 Hz, and the deflection caused by a point load of 445 kN should be below 0.4 mm.
Thermal transmittance	The varying external thermal conditions including extreme temperatures based on the systematic of EN ISO 10077-1.
Water tightness	Insulation should be predicted to prevent water flow from one floor to its lower stories. Roofs surface should have the flexibility of insulation required when green roofs are desired.
Water vapor permeability	Should be controlled by the appropriate use of water vapor barriers. Joints between floor/ roof elements themselves and with wall elements should be insulated against acoustic and vapor e.g. with rubber sealing.
Fire resistance	Ability to prevent the flow of heat and smoke according to DIN 4102. Usually 60 minutes (REI) is required for houses and apartment buildings in most countries. The joints of the floor elements to the walls shall be fire tested separately.
Acoustic insulation	The roof should have enough insulation to transfer maximum 55 db in the air, 53 db heel drops. The roof should insulate the room from any outer noises including precipitation noise.
Other insulations	Roofs should be insulated in addition to water against UV-radiations and freezing.
Durability	The floor/ roof including its connections should retain its structural integrity in design life of the building. Jointing materials must be strongly protected against corrosion e.g. by galvanization according to EN ISO 1461, EN ISO 14713 and EN ISO 10684.
Dangerous substances	Materials should not release dangerous substances in excess of the maximum permitted levels specified in European material standards and other relevant material regulations.
Transport	Dimension of elements including package and safety devices must not exceed the maximum deliverable limits defined by national transportation regulations.

Table 2-2: Product requirements of floors and roofs [84]

Property or restraint	Requirement in European Standards
Assembly	Delivering trucks should be equipped with cranes to install floor/ roof elements to their place. Therefore, appropriate handles should be predicted in elements. Implementation of required openings and conduits including openings of stair cases has to be possible.
Quality-proof production	All safety-relevant structural elements must be prefabricated in industrial workshops, while the finishing work should be done at the site by laymen or owners themselves.
Affordability	The surface finishing should be possible with local materials and unskilled labor.
Controllability	Existence and quality of decisive structural elements and connections has to be verifiable by owners by eyes or simple tools to prevent corruption.
Acceptability and attractiveness	Extra options, e.g. green roofs, should be worked out and offered. Floor surfaces should be compatible with different finishing types and materials.

Table 2-2: Product requirements of floors and roofs [84] (Continued)

2.1.3. Searching for working principles and concept variants

Working sub-structures put different functions into effect. At this stage possible working principles are searched but get not implemented, because there are still technical and cost-related unknowns, which should be found during structural and detail design phases. The same procedure has been done in chapters 5 and 6.

2.1.4. Evaluation and selection

Beside technical considerations, economic factors are of crucial importance in the design of modular systems [35]. Hence, different variants should be reviewed to select only those economically feasible. An effective way is to roughly calculate the expected production cost of each module. In addition, since basic modules appear in all sorts of variants, the most cost-efficient solutions must be selected for them. Special and adaptive modules take second position in the minimization of costs [35].

For minimizing the costs of a modular system, not only the modules themselves but also their interaction must be taken into account; in particular, the influence of special, auxiliary and adaptive modules on the cost of the basic modules [35]. For case of family houses, the effect of all types of connections should be kept in mind during design. Selecting a specific type of profile might look economically promising at the first look, but if connections are very costly in sense of production or construction, the overall price will be finally more. In addition, structures should be conceptually designed in a way that the necessary connections are mostly hinged instead of moment resisting, while providing required stability and serviceability criteria by limiting deflections and rotations. Although it looks a demanding and complex task, but the effect of price of each module on the whole price of the product must be considered.

2.1.5. Preparing dimensional layouts

After developing the principle solutions of product, the dimensional layouts can be prepared by structural engineers. This involves a stress analysis and detail design. Compared to the classical practice the work is made easier, since the engineer deals with a smaller number of variants and needs to decide between solution possibilities prepared previously for him. Another issue should be considered here is that, in design of modular systems, production and assembly considerations are of paramount economic importance to be kept in mind.

2.1.6. Preparing production documents

When the product is developed professionally, the background data and processes are documented. Product documents must be prepared in such a way that the execution of orders can be based on simple and computer-aided combinations. Once the basic modules are combined, it is easier to add extra modulus or details to the product [35]. In order to have a safe and fast assembling, also as a tool for production/construction management, both individual parts and assemblies must have mark numbers. In addition, part lists must be provided in drawings accordingly. Thanks to the high performance CAD/CAM programs like Bocad-3D and Tekla Structures, these features are deducted automatically from the 3D-model. Numbering and marking have to be related to the functionality of the module to avoid any mistake and to speed up production and assembling. This means that abbreviations used should be clear.

2.2. Modularization of products

Each product is designed to fulfill a set of different functions. It is costly to design and produce each product individually for a special function only. However, in most of the cases a rationalization is possible when the functionality is aimed by combination of definite numbers of modules. But, because modular products must have the flexibility of fulfilling several needs in different variants of the final product, they demand a greater effort in both conceptual and embody design phases.

Besides the technical and economical advantages in functionality, modular products ease production process or enable innovative automatic production lines. Each building module can be used in different products but still look individual. However, designers should have both functionality and production in design phase in mind. A large variety of functions may demand a larger number of smaller modules, but production is made more efficient with a low number of modulus. This issue will be even more advantageous where integrity plays an important role, like in structural design. Therefore an optimization is required to turn the product into a function-oriented and production-oriented one at the same time. Moreover, the position of modularization in the whole product development process is an important concern. If the modularization is started after the product is designed, there will be less flexibility rather than an early modularization. Therefore in the present work, modularization is intended from the beginning.

2.2.1. Systematic of modular products

In [35], modules are divided into function and production modules. Function modules fulfill technical functions, independently, or in combination with other modules, while production modules are considered to perform production requirements. Different types of functions, for which products are designed, are shown in figure 2-3 and explained below:

- **Basic functions:** also known as "essential modules", are the basic modules and have therefore very few variations.
- **Auxiliary functions:** joint to basic modules to add more functions to the product, but are not always needed.

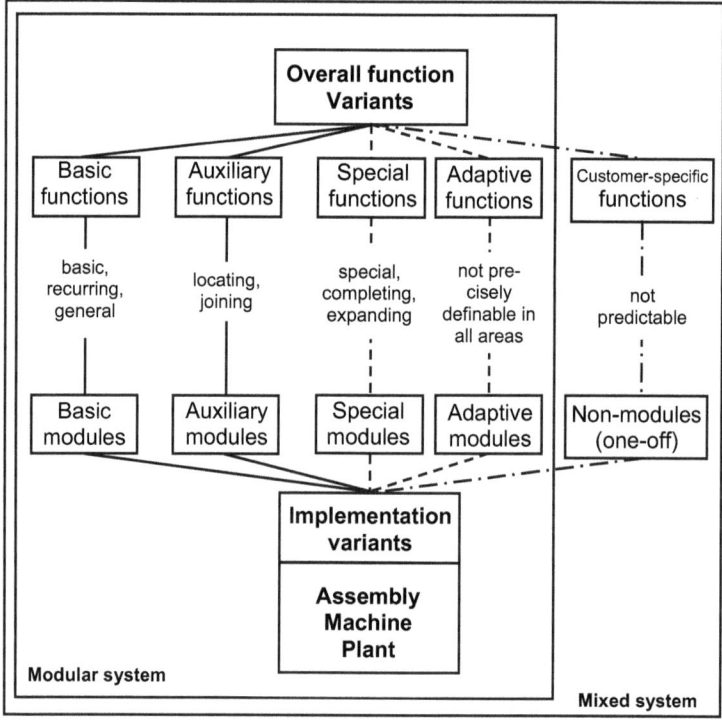

Figure 2-3: Function and module types in modular and mixed product systems [35]

- **Special functions:** also known as "possible modules", are not implemented in all variants of the product. They add one or more extra sub-functions to the product. These sub-functions bring usually added value to the product. This means, they satisfy some demands that only a minority of customers may want and pay for.

- **Adaptive functions:** are those modules without fixed dimensions, with the flexibility for unpredictable circumstances. Adaptive modules may be in turn "essential modules" or "possible modules". An example for adaptive functions is the modules which are adapted to fit the size of house exactly to the size of land in order to minimize the waste of land. This is due to the difference between total number of grids multiplied by their distance, and the original size of land.

- **Customer specific functions:** are actually non-modules that will be designed and joined to the product later on, based on the customer's order or by a more advanced product development. The final product will be a mixed system of both modules and non-modules, which can meet market and customer needs economically. An example for customer specific functions is implementation of green roofs on the roof of a residential building. To do that, the requirements have to be technically included before, to allow extra loads and isolation.

Regarding the variety of variants, modular products are divided into closed and open systems. Closed systems have a finite number of variants and their whole range and potentials can be prepared and listed in form of combinatorial schemes. Using these plans, customers can choose desired combinations directly. Open systems have an infinite number of possible combinations. In such cases, a specimen plan provides customers with typical application possibilities. For prefabricated earthquake-safe houses and schools, different details of walls, floors and roofs are closed modular systems, while arrangements of wall frames are open modular systems.

Here, one has to know the difference between components and assemblies in case of prefabricated structures. Components are those elements produced directly from cutting, boring and welding of a single piece of material, like columns or girders prepared from hot-rolled sections. Assemblies are combination of components welded or bolted together in a transportable size, like wall panels. Modularization can be implemented on both components and assemblies, but modular components are of higher advantage for cost-effective products.

2.3. Construction methods

Construction method influences the product development of buildings and has to be realized before design. Construction of prefabricated steel structures can be executed in one of the following methods [37]:

2.3.1. Platform method

In this method, the load bearing walls are installed on the floor diaphragm of the lower storey. The Vertical loads and moments are transferred to the lower storey through the diaphragm. This method is more suitable for prefabricated structures, since wall panels form directly the effective height of the storey, and also because each storey is constructed after the lower storey is finished (figure 2-4). This means that each storey is stabilized by rigid floors before the construction of next storey is started. This aspect is very important for providing safety on construction site where laymen work.

2.3.2. Balloon method

In this method, floor and roof decks are installed to the inner side of the walls and the vertical elements are continuous along the height of building (figure 2-4). In this case, special attention should be paid to accurate design and construction of floor and roof diaphragms to the main structure under horizontal loads.

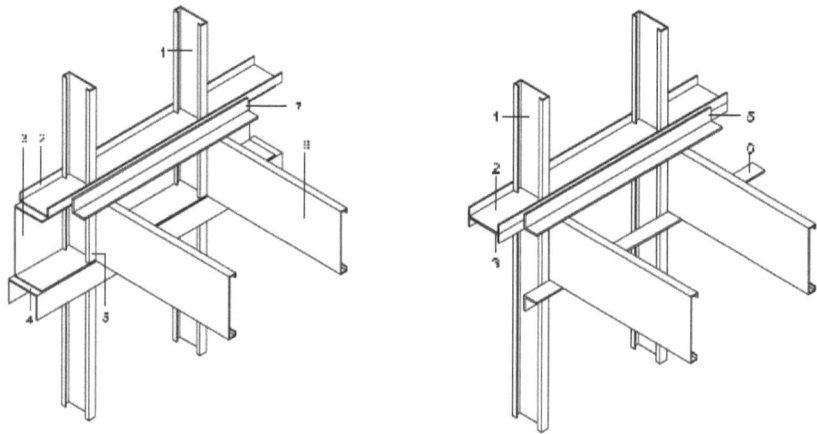

Figure 2-4: Platform (left) and balloon (right) construction methods [37]

3. Market Survey

3.1. Housing demand

The unbalance between housing demand and the capacity in construction of houses worldwide, calls for immediate and effective actions. More than one billion of the world's city residents live in inadequate housing, mostly in developing countries. Such areas are regarded as one of the most visible expressions of human poverty. The lack of adequate housing in the cities of developing countries is one of the most pressing problems of the 21st century, and the cost of providing adequate shelter for all is immense. Yet, the cost of doing nothing may be even greater, since the new urban slums are potential breeding places for social and political unrest. In figure 3-1 the geographical distribution of the housing demand is illustrated. The global housing stock in cities amounts to 700-720 million units of all types. It is estimated that 20 to 40 million urban households are homeless. A significant number of those housed, however, cannot be regarded as living in adequate shelter. Worldwide, 18 percent of all urban housing units (some 125 million units) are non-permanent structures, and 25 percent (175 million units) do not conform to building regulations. Most deficient housing units are found in the cities of developing countries, with more than half of all less-than-adequate housing units located in the Asia and Pacific region. The situation may become even worse, as household sizes decrease in most countries, and the number of urban households grows considerably faster than urban populations. In the cities of developing countries, housing delivery systems need to cope with an annual additional demand of some 18 million units, amounting to an annual increase in housing stock of nearly 5 percent [56]. As a result, the demand for affordable housing is reaching critical levels worldwide, with five million new units required per year, or 4,000 new housing alternatives needed every hour by a growing population of the poor with an unequal geographic distribution. For example, the proportion of slum dwellers to the urban population is 72% in Africa and 46% in Asia [57]. Figure 3-1 shows this distribution in different regions. The situation in rural areas of developing countries may be even worse. As rural areas are not in the focus of public and political interest, realistic data are missing. Quite often, villages in rural areas are completely destroyed in case of strong earthquakes. Having this indicator in mind, product development here explicitly includes rural areas with priority.

In order to reduce poverty worldwide, the United Nations' Millennium Development Goals addresses this issue under Goal 7, Target 4, which calls for a significant improvement in the lives of at least 100 million slum dwellers by 2020. In [58], the most effective strategy for achieving this target is mentioned to use simple and low-cost interventions.

To be able to find and develop an appropriate solution, people in need or costumers as well as their requirements have to be studied. In the first column of table 3-1, developing countries are listed according to the Human Development Report of the United Nations

Product development of earthquake-safe houses and schools

Development Programme [61]. The number of seismic events in these countries during 1974 to 2005, as well as the number of victims due to these events is also presented based on the data prepared by [62].

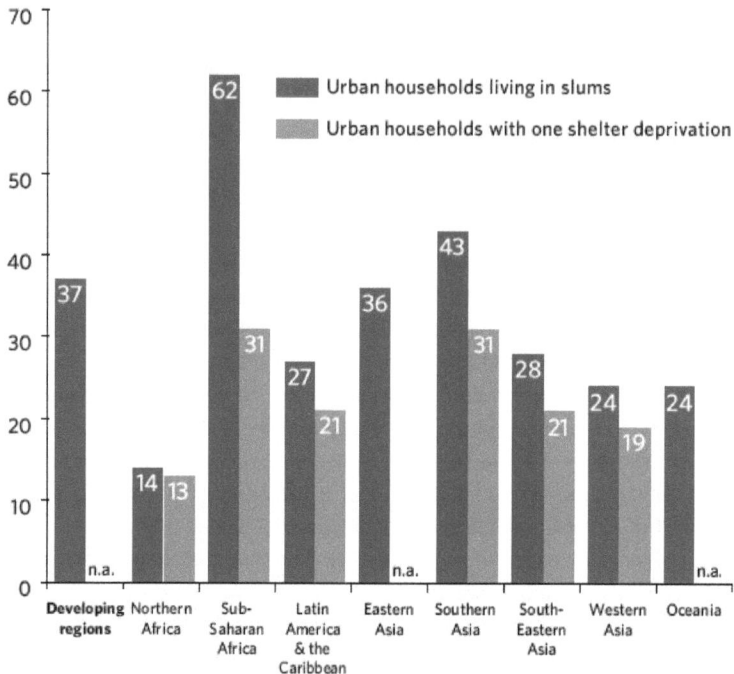

Figure 3-1: Geographical distribution of housing demand in urban regions of the world [58]

To have an estimation of the economic situation in developing countries, the Gross Domestic Production (GDP) per capita is listed in the last column of this table, using the statistics of the World Bank [63]. GDP is not the best measure for income of individuals. The income may scatter very much, especially in developing countries. But since housing projects have to be mainly financed by governments, it can be used as an estimate for the ability of that country in financing. PPP in this table refers to Purchasing Power Parity, which is a more appropriate currency converter to compare the GDP and its components across countries. In this method, an international dollar has the same purchasing power over Gross National Income (GNI) as a U.S. dollar has in the United States [64].

Country	Total number of seismic disasters (1974-2007)	Total number of victims (killed and affected)	Purchasing power parity (PPP) (international dollars)
India	18	27.916.982	2.740
China	91	21.415.092	5.370
Pakistan	20	6.573.780	2.570
Indonesia	77	6.307.159	3.580
Turkey	39	6.221.323	12.350
Guatemala	10	5.051.959	4.520
Mexico	20	2.561.786	12.580
El Salvador	6	2.412.551	5.640
Iran (Islamic Republic of)	70	2.387.765	10.800
Philippines	13	2.228.792	3.730
Chile	8	1.596.865	12.590
Colombia	16	1.326.716	6.640
Algeria	14	1.192.469	7.640
Peru	24	1.139.377	7.240
Sri Lanka	1	1.054.705	4.210
Afghanistan	25	628.672	--
Nepal	3	542.710	1.040
Yemen Arab Rep	1	403.007	2.200
Ecuador	10	215.258	7.040
Taiwan (China)	5	111.790	--
Somalia	1	105.381	--
Egypt	3	93.530	5.400
Thailand	2	75.352	7.880
Uganda	4	52.007	920
Malawi	1	50.109	750
Papua New Guinea	12	48.191	1.870
Argentina	3	46.141	12.990
Yemen	1	40.049	--
Costa Rica	10	31.441	10.700
Maldives	1	27.316	5.040
Brazil	2	23.288	9.370
Guinea	1	21.711	1.120
Panama	4	21.543	10.610
Bangladesh	7	19.161	1.340
Bolivia	3	18.170	4.140
Myanmar	3	15.931	--
Vanuatu	9	15.217	3.410
Morocco	1	14.093	3.990
Nicaragua	4	14.080	2.520
Tanzania (United Republic of)	5	0.500	1.200
Sudan	2	8.018	1.880
Venezuela (Bolivarian Republic of)	6	7.467	11.920

Table 3-1: Number of seismic disasters and victims during 1974-2007 and the level of GDP in developing countries [62, 63]

Country	Total number of seismic disasters (1974-2007)	Total number of victims (killed and affected)	Purchasing power parity (PPP) (international dollars)
Cuba	2	5.878	--
Tonga	1	5.506	3.650
Malaysia	1	5.143	13.570
Seychelles	1	4.833	15.450
Solomon Islands	4	4.171	1.680
Honduras	4	4.105	3.620
Dominican Republic	1	2.018	6.340
Cyprus	1	1.867	26.370
Rwanda	1	1.688	860
Congo	1	1.511	2.750
Mozambique	1	1.480	690
Iraq	1	520	--
South Africa	6	509	9.560
Congo (Dem. Rep. of the)	1	320	290
Burundi	1	123	330
Dominica	1	100	7.410
Trinidad and Tobago	1	17	22.490
Barbados	1	1	16.140
Kenya	2	1	1.540
Fiji	2	0	4.370
Korea (Democratic People's Republic of)	1	0	24.750
Saint Lucia	1	0	9.430
Samoa	1	0	3.930

Table 3-1: Number of seismic disasters and victims during 1974-2007 and the level of GDP in developing countries [62, 63] (Continued)

Under term costumers, the following two groups of people are meant at this work:

The first time home owners: this group includes those households who are going to live independently under a shelter, or those who currently suffer from a non-adequate or low-standard shelter. While the former group may be able to buy the house on their own, the latter needs extra financial assistance.

The disaster survivals: who have got homeless as result of a disaster and are in urgent need of a house. Post disaster reconstruction of the affected areas and accommodation of the survivals requires both financial and technical efforts in a public scale.

3.2. Demand on schools

In the same context mentioned above and from an objective point of view, schools posses also a high priority. In many regions of the world, schools are constructed with the same effort and techniques as houses, and are threatened therefore by seismic hazard. These may endangers their occupants and interrupt or destroy their important functions. In addition, schools and other larger structures often serve as community shelters in time of need, so their loss is a double burden on an afflicted locality [59]. Citing Pakistan government estimates, UNICEF has stated that at least 17,000 school children died when 6,700 schools were destroyed during the earthquake in 8 October 2005 in North-West Frontier Province and 1,300 in Pakistan-administered Kashmir as children attended morning classes [60]. Also in the earthquake of 12 May 2008 in Sichuan province of China 19,065 students died according to the Chinese government statement during destruction of 7,000 classrooms [69]. Based on the estimation presented in [59], roughly one billion children aged 0-14 live in countries with a high seismic risk. Several hundred million are at risk when they are attending school. Therefore, a huge number of schools have to be accomplished annually, to fulfill this demand. Table 3-2 shows the primary school population in seismic hazard in top 20 countries for earthquake fatalities from 1900 to 2000, stated in [59]. Column "children out of school who should attend" in this table includes those children in rural areas with poor infrastructure.

Nation	Age Group	School-age population	Children out of school who should attend
China	7-11	110,499,000	8,054,600
Japan	6-11	7,335,000	300,000
Italy	6-10	2,789,000	6,400
Iran	6-10	9,221,000	2,436,300
Turkey	6-11	7,969,000	no data
Peru	6-11	3,416,000	4,600
Armenia	7-9	199,000	no data
Pakistan	5-9	19,535,000	7,785,400
Indonesia	7-12	26,081,000	2,046,300
Chile	6-11	1,751,000	1,956,000
India	6-10	112,469,000	no data
Venezuela	6-11	3,286,000	394,600
Guatemala	7-12	1,869,000	293,300
Afghanistan	7-12	3,372,000	no data
Mexico	6-11	13,070,000	78,400
Nicaragua	7-12	810,000	155,900
Morocco	6-11	4,071,000	8,952,000
Nepal	6-10	3,065,000	846,800
Taiwan	no data	no data	no data
Philippines	6-11	11,330,000	822,600

Table 3-2: Primary Education data on top 20 countries for earthquake fatalities 1900-2000 (in order) [59]

3.3. Seismic Risk

The potential economic, social and environmental consequences of hazardous events, in form of damage or fatality that may occur in a specified period of time are defined by seismic risk. The level of seismic risk is a function of the factors that determine the potential for people to be exposed to this type of natural hazard. In other words, to understand earthquake risk, both seismic hazard and also the different types of vulnerability to that hazard have to be known. Seismic risk should be understood in the context of socio-economic

Product development of earthquake-safe houses and schools

systems, which determine people's health, income, building safety and access to information at the time of the disaster, and the effectiveness of response [65].

There are several definitions of the seismic risk available in the literature. In a comprehensive definition in [65], risk is defined in terms of the seismic hazard itself; the vulnerability of the built environment due to structural weaknesses and seismic resistance; the human, socio-economic and cultural impact of an earthquake (value); and the country's preparedness and response to an event (management). Therefore:

$$\text{Earthquake Risk} = \frac{\text{Seismic Hazard} \times \text{Vulnerability} \times \text{Value}}{\text{Management}}$$

By weighting and combination of the corresponding parameters, different risk maps can be prepared for different purposes. Figures 3-2 and 3-3 show two global maps prepared by the World Bank showing the proportion of population and GDP in at least one hazard for different countries of the world, respectively. These hazards and their rate of occurrence in developing countries are shown in figure 3-4, cited by International Strategy for Disaster Reduction (ISDR) based on the data of Emergency Events Database (EM-DAT) [62]. As it can be seen in this graph, 64% of the whole hazards in developing countries are earthquakes and tsunamis.

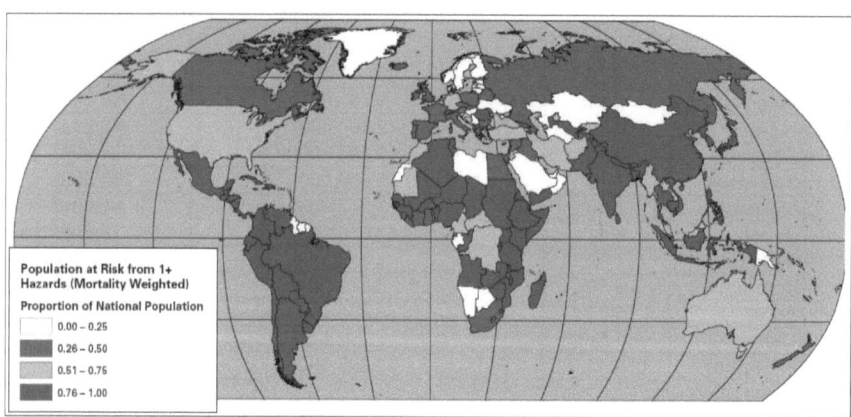

Figure 3-2: Proportion of population in at least one hazard for different countries of the world [80]

3. Market survey

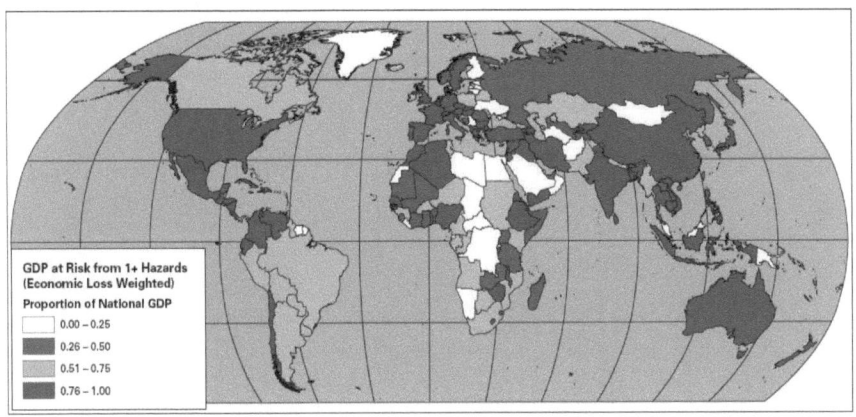

Figure 3-3: Proportion of GDP in at least one hazard for different countries of the world [80]

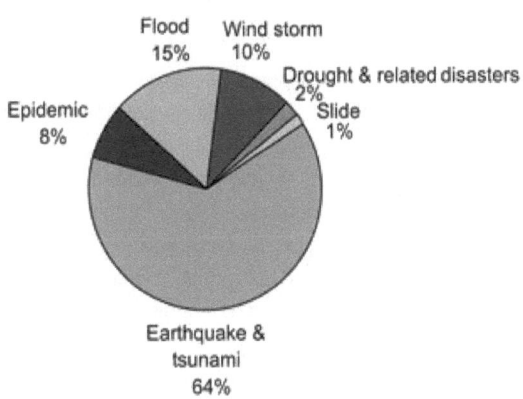

Figure 3-4: Proportion of people killed due to different natural hazards in developing countries 1991-2005 [68]

3.3.1. Seismic Hazard

For all construction activities, including new construction or post-disaster reconstruction, regional seismic hazard has to be well understood. While underestimating hazard leads to an unacceptable level of risk, overestimation results to an economical unaffordable solution. In figure 3-5 a global seismic hazard map is shown, which was prepared during the Global Seismic Hazard Assessment Program (GSHAP) in 1999 [66]. As it is obvious from this map, seismic hazard is very high in many developing countries in the Middle East and East Asia as well as in Latin America.

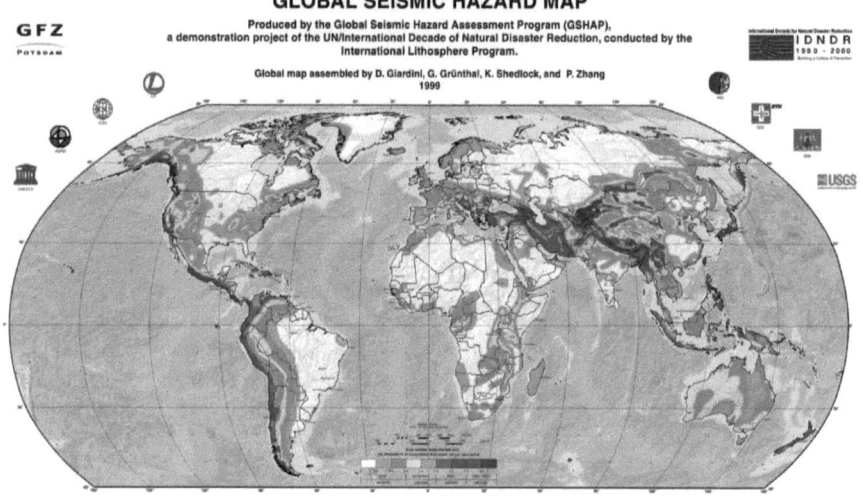

Figure 3-5: Geographical distribution of seismic hazard in the world [66]

Reliable and realistic consideration of seismic hazard is the key element of efficient risk mitigation. This consideration needs analysis of seismic hazard, which is usually done either deterministically, for a particular earthquake event, or probabilistically, for probable future earthquakes. Since the probabilistic analysis includes the uncertainty of future earthquakes (i.e. uncertainty in time, location and magnitude of the event), it is used broadly for seismic hazard analysis purposes. Inputs of this study known as Probabilistic Seismic Hazard Analysis (PSHA) include: (i) earth science, which is defined by specifying seismicity and geometry of seismic source zones expected to be the source of future earthquakes, and (ii) ground motion information, which is defined by an attenuation relationship that estimates ground motion parameters from earthquake magnitude and source-to-site distance for various site conditions.

Results of a PSHA are used to prepare hazard maps in macroseismic or microseismic scales and for special site studies. These hazard maps, which show the contribution of seismic hazard parameters, are very important for almost all seismic hazard mitigations. In fact, hazard maps are a visual presentation of the seismic hazard level in different regions and are the basis of seismic design codes and regulations. These maps are usually prepared as a function of different ground motion parameters, i.e. Peak Ground Acceleration (PGA), Peak Ground Velocity (PGV), Peak Ground Displacement (PGD) and Spectral Acceleration (SA) in short periods, normally 0.2 sec and long periods, normally 1.0 sec. These periods comply with those of high-rise and low-rise buildings, respectively. They are also prepared for different probabilities of exceedance in a reference return period. The reason for such a variety, rather than just PGA maps in traditional approach, is various characteristics that these maps introduce. Nevertheless, hazard maps should be updated as new events decrease the uncertainty, e.g. clarify the existence of unknown faults, or as new scientific approaches introduce new methods. However, preparation of such maps on national scale is time-consuming and as a result it will be decisive to find out the regions with the highest priority. Therefore, a division of the country based on the hazard status will be more efficient [19]. Such divisions indicate more accurately both the priority of each region and the type of rehabilitation procedure, which facilitate an affordable earthquake resistant housing programme.

High toll of life losses and extensive destruction due to recurring earthquakes demand serious and immediate hazard mitigation plans. As an example for this high recurrence of the events, the peak ground acceleration-frequency curve for Iran is illustrated in figure 3-6. Relatively high frequency of events in this curve reveals that all construction activities must be planned and designed considering earthquake effects.

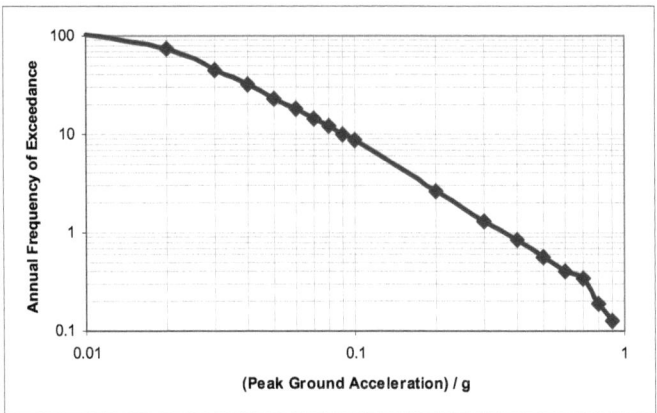

Figure 3-6: Peak acceleration-frequency of Iran, data 1973 to 2004

3.3.2. Seismic vulnerability

High number of fatalities and wide spread damages, especially in developing countries, during last earthquakes show, that houses in many regions of the world possess a high seismic vulnerability due to inappropriate construction techniques.

Earthquake vulnerability is determined by the probability of failure of the built environment due to the natural hazard. Vulnerability of a building shows how it responds to an earthquake. In order to measure vulnerability for engineering purposes, EMS-98 [67] presents vulnerability classes A to F and assigns them to different building types. This assignment is shown in table 3-3, where the most likely vulnerability of each type is shown as an interval. It is notable to mention that in this table class A refers to a highly vulnerable building, while class F represents a very high level of earthquake-resistance.

Product development of earthquake-safe houses and schools

Type of Structure		Vulnerability Class A B C D E F
MASONRY	rubble stone, fieldstone	O
	adobe (earth brick)	O⊢
	simple stone	⊢·O
	massive stone	⊢O⊣
	unreinforced, with manufactured stone units	⊢·O·⊣
	unreinforced, with RC floors	⊢O·⊣
	reinforced or confined	⊢·O⊣
REINFORCED CONCRETE (RC)	frame without earthquake-resistant design (ERD)	⊢···O·⊣
	frame with moderate level of ERD	⊢···O⊣
	frame with high level of ERD	⊢···O⊣
	walls without ERD	⊢·O⊣
	walls with moderate level of ERD	⊢··O⊣
	walls with high level of ERD	⊢··O⊣
STEEL	steel structures	⊢····O⊣
WOOD	timber structures	⊢···O⊣

O most likely vulnerability class; — probable range; ····· range of less probable, exceptional cases

Table 3-3: Classification of buildings into different vulnerability classes (Vulnerability Table) [67]

According to EMS-98, besides the structural type, the following factors can influence vulnerability of a building type and have to be contemplated while evaluating or aiming at reducing the vulnerability:

Quality and workmanship: A poor quality of construction can increase the vulnerability to a great deal. Poor quality may be due to carelessness or cost-cutting measures both can be observed in developing countries. One of the solutions for this problem is to reduce the effect of workmanship on structural safety by distinguishing between the structural-relevant parts produced in an industrial way, and secondary parts, e.g. infill walls, accomplished by local labor.

State of preservation: This depends to the level of maintenance of the building. Buildings which are allowed to decay may lose their stiffness and the vulnerability class can be reduced for at least one class. This is also true for buildings affected by an earthquake or a foreshock; these buildings usually get weaker and their level of maintenance must be evaluated by a structural expert.

Regularity: In order to have a more predictable and reliable performance of the building under earthquake loads, the building has to be as symmetric as possible. Symmetry should be considered in its most general meaning, including symmetry in dimensions of plan and elevation, and symmetry in distribution of lateral load bearing components. Some collapse examples in chapter 4 show consequences of irregularity.

Ductility: Ductility is a measure of a building's ability to withstand lateral loading in a post elastic range, i.e. by dissipating earthquake energy and creating unimportant damage in a controlled wide spread or locally concentrated manner, depending on the construction type and structural system [67]. While some structural types, mainly steel structures possess a high ductility, some others, like adobe and masonry are brittle and collapse during shaking. This phenomenon is discussed in detail in chapter 5.

Position: Buildings located in an unfavorable arrangement can affect other buildings or be affected by them. Hence, the vulnerability will be increased unexpectedly if earthquake safety is not considered in urban planning.

Strengthening: Seismic vulnerability of buildings can be modified by strengthening and retrofitting measures. However, the functionality of these measures should be proved and they should fulfill economic criteria.

Earthquake resistant design (ERD): As mentioned before, seismic hazard differs from a region to another. To design a building for a special region, its regional requirements should be studied. This has been usually considered in national codes of each country in form of zoning maps. While it is not economic to imply the highest level of earthquake resistant design to buildings for all regions, the level of seismic hazard should be conservatively interpreted for different regions in order to reduce the seismic vulnerability.

3.4. Major construction types in developing countries

As mentioned, the level of risk depends on the seismic vulnerability. The construction type defines the vulnerability class of the building. Therefore, the main construction types in developing countries are presented and discussed here. Descriptions of major types are adopted from World Housing Encyclopedia [9]. More details regarding each region can be found there.

3.4.1. Adobe houses

Adobe mud blocks are one of the oldest and most widely used building materials. The use of adobe is very common in some of the world's most hazard-prone regions, such as Latin America, Africa, the Indian subcontinent and other parts of Asia, the Middle East (figure 3-7), and southern Europe. Around 30% of the world's population lives in earth-made construction. Approximately 50% of the population in developing countries, including the majority of the rural population and at least 20% of the urban and suburban population, live in earthen dwellings. Generally, low-income rural populations use this type of construction.

Figure 3-7: Zavareh, a small town in central in Iran [19] and a typical rural adobe house in Iran [9]

Adobe is a low-cost, readily available construction material, usually manufactured by local communities. Adobe structures are generally self-made because the construction practice is simple and does not require additional energy resources. Often the blocks are made from local soil in a homeowner's yard or nearby. Mud mortar is typically used between the blocks (figure 3-8). Worldwide use of adobe is mainly in rural areas, where houses are typically one story, 3.0 m high, with wall thicknesses ranging from 0.25 m to 0.80 m. Urban adobe houses are found in most developing countries. In Latin America, adobe is mainly used by low-income families, whereas in the Middle East (e.g., Iran), it is used both by wealthy families in luxurious residences as well as by poor families in modest houses.

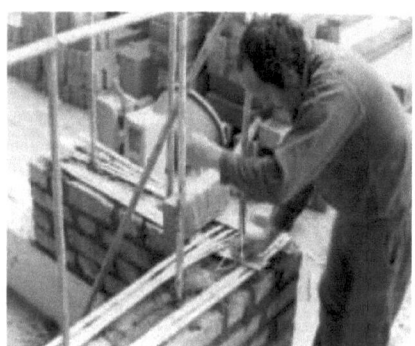 

Figure 3-8: Construction with adobe blocks and mortar [81]

Architectural characteristics are similar in most countries: the rectangular plan, single door, and small lateral windows are predominant. Quality of construction in urban areas is generally superior to that in rural areas. The foundation, if present, is made of medium-to-large stones joined with mud or coarse mortar. Walls are made with adobe blocks joined with mud mortar. Sometimes straw or wheat husk is added to the soil used to make the blocks and mortar. The size of adobe blocks varies from region to region. In traditional constructions, wall thickness depends on the weather conditions of the region. Thus, in coastal areas with a mild climate, walls are thinner than in the cold highlands or in the hottest deserts. The roof is made of wood joists (usually from locally available tree trunks) resting directly on the walls or supported inside indentations on top of the walls. Roof covering may be corrugated zinc sheets or clay tiles, depending on the economic situation of the owner and the cultural preferences of the region.

In addition to its low cost and simple construction technology, adobe construction has other advantages, such as excellent thermal and acoustic properties. However, most traditional pure adobe construction responds very poorly to earthquake ground shaking, suffering serious structural damage or collapse and causing a significant loss of life and property, which have been addressed by WHE [9]. Adobe buildings are not safe in seismic areas because their walls are heavy and they have low strength and brittle behavior. During strong earthquakes, due to their large mass, these structures develop high levels of seismic forces, which they are unable to resist, and therefore they fail abruptly. Typical modes of failure during earthquakes are severe cracking and disintegration of walls, separation of walls at the corners, and separation of roofs from the walls, which can lead to collapse. A typical adobe buildings collapsed are shown in figure 3-9.

Figure 3-9:   Typical adobe buildings destroyed during earthquake in Bam, Iran [Courtesy: Astan-Qods-Razavi]

3.4.2. Wood houses

Wood construction is common for many single-family houses throughout the world. In areas where timber and wood materials are easily accessible, wood construction is often considered to be the cheapest and best approach for small housing structures. Wrong connection of the building to its foundation is a critical issue for all types of wood structures. Using weak wooden substructures as the basement of the structure in some rural areas shown in figure 3-10 is observed to be a cause of vulnerability [19].

Figure 3-10: Wooden base and foundation construction in a rural area of Gilan province in Iran [19]

Inadequate shear resistance may also be a problem for structures that are deficient in sheathing or bracing. Inappropriate or poor constructed connections in many traditional wooden structures decrease the seismic performance of these buildings. Inadequate blocking in walls and between joists in the roof and floor diaphragms for frame structures, especially at stress concentrations, can be a significant building weakness. Lack of proper building maintenance, like in figure 3-10, is another issue for all timber buildings.

3.4.3. Stone masonry houses

Stone masonry houses are used widely worldwide. The main material used in the walls is blocks of available natural stone. There are stone masonry houses with and without mortar. Mortar is either mud-based or cement-based. A variety of roofing systems are adopted including tiled roof supported on wood trusses, asbestos or steel sheets on steel trusses, and reinforced concrete slab. The main lateral and gravity load-resisting system consists of stone masonry structural walls. The walls are generally uniformly distributed in both orthogonal directions with a wall thickness ranging from 400 mm to 700 mm. The wall density (area of walls in one direction versus total plan area) ranges from 5% to 25%.

Stone masonry is a traditional form of construction practiced for centuries in the regions where stone is locally available. It is still found in old historic centers, often in buildings of cultural and historical significance, and in developing countries where it represents affordable and cost-effective housing construction. This construction type is present in earthquake-prone regions of the world, such as Mediterranean Europe and North Africa, the Middle East, India, Nepal, and other parts of Asia. Houses are built by local builders or by owners themselves without any formal training. The quality of construction in urban areas is generally superior to that found in rural areas.

Structural walls are supported either by stone masonry strip footings or there are no footings at all. Floor structures in towns and historic centers are vaulted brick masonry at the ground floor level and timber joists at the upper floor levels. Timber joists are usually placed on walls without any physical connection. The original floor structures in historic buildings have typically been replaced either by a pre-cast joist system or by solid reinforced concrete slabs.

The most important factors affecting the seismic performance of these buildings are:

- The strength of the stone and mortar
- The quality of construction
- The density and distribution of structural walls
- Wall intersections and floor/roof-wall connections

Stone masonry construction generally shows very poor seismic performance shown in figure 3-11. Poor quality of mortar is the main reason for the low tensile strength of rubble stone masonry. Timber floor and roof structures are usually not heavy and therefore do not induce large seismic forces. However, typical timber floor structures are made of timber joists that are not properly connected to structural walls. These structures are rather flexible and are not able to act as rigid diaphragms. Due to their large thickness, stone masonry walls are rather heavy and induce significant seismic forces.

Out-of-plane failure can occur when the connections between the exterior and interior walls are inadequate. When the connections between the perpendicular walls are strong, the wall shear capacity can be exhausted, thus causing typical shear cracks to develop. Brittle shear walls are the weak point of this design concept.

Figure 3-11: A typical stone building in Iran and a damage pattern in external walls during earthquake [9]

3.4.4. Brick masonry houses

Brick masonry houses are another common construction type. Clay mud is used to form regular-sized masonry units. These units are sometimes burnt in a kiln, or simply sundried. Brick masonry houses are made with and without mortar either mud-based or cement-based. These units are the main materials used in the walls. Again, a variety of roofing systems are adopted including tiled roof supported on wood trusses, asbestos or steel sheets on steel trusses, and reinforced concrete slab. According to the EMS scale, brick buildings fall generally in Class B, except for the examples of modern design or buildings with seismic strengthening, which have been classified as Class C.

Common damage patterns in residential buildings reported in WHE include the following:

- Shear cracks in the walls, mainly starting from corners of openings (MMI intensity VIII)
- Partial or complete out-of-plane wall collapse due to lack of wall-to-wall anchorage and wall-to-roof anchorage. In extreme cases this is accompanied by partial or total collapse of floor and roof structures (MMI intensity VIII-IX)
- Total collapse of walls as shown in figure 3-12 and entire buildings in some cases (MMI intensity X)

Figure 3-12:   Damage to brick masonry buildings in earthquake, Iran [31]

3.4.5. Confined masonry houses

This type of housing is practiced in many vernacular forms worldwide, particularly along the Alpine-Himalayan belt (figure 3-13). These are load bearing masonry houses improved with the help of wood or concrete frame members introduced in the walls to reduce the masonry walls into smaller panels that are more capable of withstanding earthquake shaking. The masonry could be made by either stone or brick. This system is superior to the traditional load-bearing masonry houses, because of shear wall effect. A variety of roofing systems are employed with the confined masonry wall system, depending on the geographic region of construction.

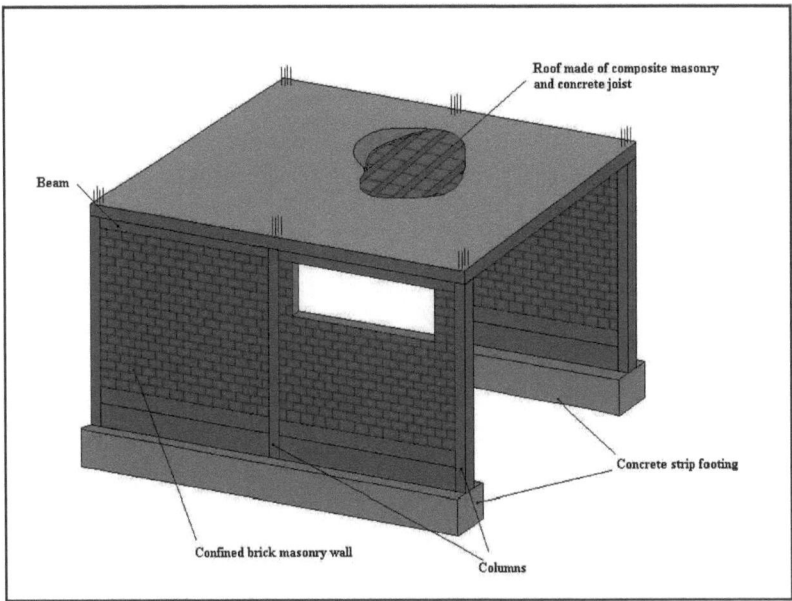

Figure 3-13: The main elements of a typical confined-masonry building [9]

Confined masonry construction consists of un-reinforced masonry walls confined with reinforced concrete (RC) tie-columns and RC tie-beams. The tie-columns and tie-beams provide confinement in the plane of the walls and also reduce out-of-plane bending effects in the walls. The walls are made of different masonry units, ranging from hollow clay or hollow concrete blocks to solid masonry units of either clay or concrete. This type of construction is used both in urban and rural areas, either for single-family residential construction or for multifamily construction up to four or five stories.

3.4.6. Reinforced concrete frame buildings

This type of housing is becoming increasingly popular across the world, particularly for urban construction. It employs beams (i.e., long horizontal members), columns (i.e., slender vertical members) and slabs (i.e., plate-like flat members), to form the basic structure for carrying the loads. Vertical walls made of masonry or other materials are used to fill in between the beam-column grids to make functional spaces. These houses are expected to be constructed based on engineering calculations. However, in a large part of the developing world, such buildings are being built with little or no engineering calculations (figure 3-14). These structures are cast monolithically, that is, beams and columns are cast in a single operation in order to act in unison. RC frames provide resistance to both gravity and lateral loads through bending in beams and columns if properly designed, detailed and constructed. There are several subtypes of RC frame construction:

- Non-ductile RC frames with/without infill walls
- Non-ductile RC frames with reinforced infill walls
- Ductile RC frames with/without infill walls

In many regions of developing countries, reinforced concrete buildings do not possess a ductile design. Thus, non-ductile concrete frames, although often designed to resist lateral forces, do not incorporate modern ductile seismic detailing provisions.

Figure 3-14: Damage to reinforced concrete buildings due to inadequate horizontal load resisting systems in Ardakul earthquake, Iran [31]

In several instances, seismic performance of RC frame buildings has been quite poor, even when subjected to earthquakes below the design level prescribed by the code. The key deficiencies identified in the RC frame construction practice include the following:

- Alteration of the member sizes during the construction phase from specifications in the design drawings
- Noncompliance of the detailing work with the design drawings
- Inferior quality of building materials and improper concrete-mix design

- Modifications in the structural system performed by adding/removing components without engineering input (figure 3-15)
- Reduction in the amount of steel reinforcement as compared to the design specifications
- Poor construction practice

Figure 3-15: Reducing the column cross-section to connect door to reinforcing bars

4. Last earthquakes: costs paid, lessons learnt

In this chapter the experiences gathered from former earthquakes are categorized and explained. The importance of this categorization is considerable from two points of view:

1) Earthquakes happen in very different manners and earthquake engineering has therefore no definite single solution, which can be used as a faultless rule for earthquake-resistant design and construction of buildings. All developed methods are rather based on the former events. For this reason, consequences of each event can be used for modification and development of the existing and new buildings, if they are precisely investigated and studied.

2) Each failure is an experiment, which shows causes of the collapse. Once the causes are known, engineers can avoid the failure occurrence by eliminating its causes in future by continuous product development. Earthquake engineers can learn from the failures and from the success, i.e. houses surviving earthquakes when others did not (figure 4-1).

Figure 4-1:   Collapse of a school building in Sichuan, China [13]

4.1. Architectural related aspects

4.1.1. Collaboration of the architect and the engineer

The concept of earthquake-resilience should be taken in mind from the early stages of planning a building. If a building is not appropriately planned, its design for earthquake-resistance is either impossible or very costly. Currently, planning process of houses and schools starts in most of the cases with an architectural plan without taking the earthquake-resistance rules in mind. Usually, people engaged in the supply network of planning and design, are not aware of their contribution in earthquake safety, and believe that this criterion can be satisfied later on by an appropriate seismic design done by a structural engineer. This problem is addressed in many related literatures and effort is called for to initiate engagement of architects, authorities and project owners to increase the seismic safety of the buildings. Figure 4-1 shows the collapse of a school building in Sichuan, China while its neighbor building has survived the earthquake.

4.1.2. Simple solutions instead of complicated ones

In general, simple solutions are preferred, since they are less prone to the errors or mistakes, which can be created in turn intentionally or unintentionally. This fact is historically proven for instance in case of half-timber structures. In figure 4-2 a traditional house with diagonal bracings is shown, which withstood an earthquake of magnitude 5.9 on the Richter scale in Turkey.

Figure 4-2: A traditional braced house withstood an earthquake of magnitude 5.9 on the Richter Scale in Yuva Village, Cankiri Province, Turkey [14]

Having inclined columns is another common complexity, which can increase the vulnerability of the structure due to the complex distribution and interaction of forces in the column. Inclined columns should be designed and detailed for normal, bending and shear (P-M-V) stresses instead of for only normal and bending (P-M) stresses.

Another category of examples are large cantilever beams or plates observed in many buildings. They are very vulnerable to the earthquake forces, since there is usually more vibration in the cantilever beams and plates. This is the reason why codes consider the vertical component of the earthquake for design of the cantilever beams. Figure 4-3 shows a collapse happened in a suspended roof during an earthquake in Chile.

Figure 4-3: Roof collapse in the city of Antofagasta, Chile [15]

High buildings should also be substituted by simple but safer low-rise buildings, wherever possible, because higher buildings are threatened largely by seismic hazard. As shown in figure 4-4, tall buildings can tilt over as a matter of their geometry, since the overturning moment is relatively large in comparison with the resisting moment. As shown in figure 4-5, overturning moment, M_O, is product of horizontal forces by the distance from the center of gravity to the ground level, while resisting moment, M_R, is defined as product of the effective weight by minimum distance from centre of gravity in the plan to the foundation-edge.

In addition, the probability of appearance of higher modes of vibration is higher in tall buildings. The taller the building, the higher is the number of contributing modes. This effect should be thoroughly considered in case of high-rise buildings, which demands more time and costs.

Figure 4-4: Turn over of slim high-rise buildings, Niigata, Japan [16]

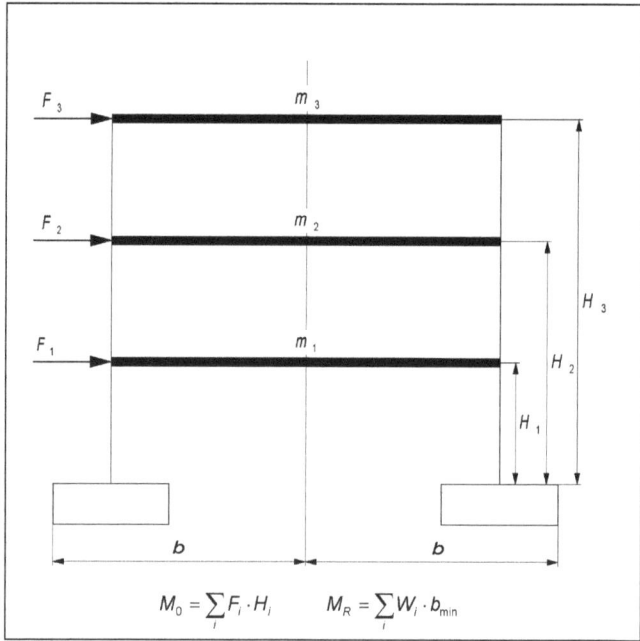

Figure 4-5: Seismic induced overturning and resisting moments

4.1.3. Symmetry and compactness

In buildings with wide or asymmetric architectural layout plans, besides the first transitional mode, rotational and higher modes can be activated during an earthquake. As mentioned before, in conventional practice buildings are not calculated for effect of these modes, which may exceed the capacity of the structure. Therefore, it is advised to favor compact plan configurations. Figure 4-6 shows a failure because of the torsion in the building. Besides the geometry, asymmetric distribution of lateral load bearing elements, i.e. bracings and shear walls, leads to this deficiency. Asymmetry should be avoided also in the elevation. This includes any sudden decrease in lateral stiffness of the storey, by changing the lateral load bearing systems. Therefore, the lateral load-bearing elements, e.g. bracings, should be distributed over the height of the building.

Figure 4-6: Collapse due to asymmetry in the plan, Yalova, Turkey [12]

4.1.4. Soft-storey floors

There is a common tendency in planning, especially in developing countries, to have no walls or to have large spans without any shear walls in the bottom floor of the buildings, for instance to use the base floor fully as shops or garages. Consequently, bracings or walls are eliminated in such cases. This makes the first storey softer than the upper ones and causes development of plastic hinges at the bottom or top of columns in the first floor. Collapse of bottom floors initiate the failure of the whole building. Basically, the columns of the bottom storey have the largest stresses, since they bear the weight of the whole building as well as the soil-structure interaction. Therefore, they should not be used as energy dissipation zones [8]. Figure 4-7 shows collapse due to the soft-storey effect in a school building in Algeria and in a house in Iran during the earthquake.

The soft-storey effect can also happen in upper stories of a building, where the lateral stiffness is suddenly lower in a storey than its upper ones. This happens usually in case of increasing the span or elimination of some braces in a specific storey.

Product development of earthquake-safe houses and schools

Figure 4-7: Soft-storey ground floor of a school, Boumerdes, Algeria [11], and of a house in Bam, Iran [Courtesy: Astan-Qods-Razavi]

4.1.5. Openings

As can be seen in figure 4-8, large amount or large size of the openings is a cause of stress concentration and initiation of cracks. This fact should be considered in planning of the building and during the life cycle of the building. In architectural design of buildings, short columns, which are usually created in columns adjacent to the openings, should also be avoided. Short columns are those columns that are relatively thick compared to their height, and are often fixed in strong beams or slabs. The shear failure of short columns is a frequent cause of collapse during earthquakes. Short columns may be created unintentionally by inappropriate addition of infill walls in the frame structures. Columns under horizontal actions in frame structures may be stressed up to their plastic moment capacity. In the case of short columns with considerable bending capacity, an enormous moment gradient and thus a large shear force results. This often leads to a shear failure before reaching the plastic moment capacity [8].

Figure 4-8: Collapse of a wall with large openings, Yalova, Turkey [12]

4.2. Structural engineering aspects

4.2.1. Seismic provisions of the codes

One of the obvious lessons of last earthquakes is that the pre-engineered buildings designed and constructed according to the codes perform better in case of earthquake, rather than those not designed for earthquake effects like in figure 4-9. This matter can be better observed in the developing countries in which both categories of pre-engineered and non-engineered buildings are constructed over a relatively short period of time. As an example one can compare this behavior in Iran. The first seismic code of Iran was prepared after a devastating earthquake of September 1st 1962 in Buin-Zahra with magnitude of 7.2 on the Richter scale, which caused 12,000 casualties. Efficiency of this code was approved after the earthquake of 1990 in Rudbar-Manjil and made it popular. In all buildings constructed with permission from the municipalities, this code was more or less conserved. The code was developed continuously in three editions, adopting other international codes. Consequently, Iranian buildings can be categorized into three groups, which are constructed according one of the three editions of the code in 1988, 1999, or 2005. Table 4-1 shows the time of construction of existing buildings in Iran [19].

Figure 4-9: Elementary school collapsed because of insufficient lateral resisting system and soft-storey, Boumerdes, Algeria [11]

Another important issue is that the codes apply for each case a safety factor, which depends on the uncertainties and probability of diversion from the assumptions taken to prepare the code, as well as on the structural reliability. Hence, these safety factors should be strictly respected and implemented in all practices rather than be eliminated or reduced, even if "it looks too high". In this regard, product development can help systematically that codes are respected.

The year of construction completion	1986 Census			1996 Census		
	Total country	Urban areas	Rural areas	Total country	Urban areas	Rural areas
Total	8217375	4669722	3547653	10770112	6913730	3856382
	100.0%	56.8%	43.2%	131.1%	84.1%	46.9%
1996	-	-	-	205679	143046	62633
1986-1995	[1]215482	[1]134557	[1]80925	4043511	2469958	1573553
1976-1985	4551277	2638320	1912957	3640011	2554950	1085061
1966-1975	1609207	1010391	598816	1454396	946145	508251
Prior to 1966	1727041	815400	911641	1317117	727305	589812
Not stated	114368	71054	43314	109398	72326	37072

1. Only for the year 1986.

Table 4-1: Conventional housing units in urban and rural areas by the year of construction completion [19]

4.2.2. Site specific response spectra

In order to prepare seismic regulations for each region, local relative seismic hazard has to be studied thoroughly in each region. Hence, the seismicity of that region is investigated, utilizing a map of active faults, earthquake catalogues, and attenuation laws of the seismic waves in that region. The results are presented as hazard maps and design spectrums in the code. The latter shown in figure 4-10 includes the effect of possible earthquakes in the region with a high level of certainty. This spectrum relates the natural frequency or natural period of equivalent systems with one degree of freedom to the seismic action applied to the building. When designing for a region, the correct coefficients have to be taken for analysis and design.

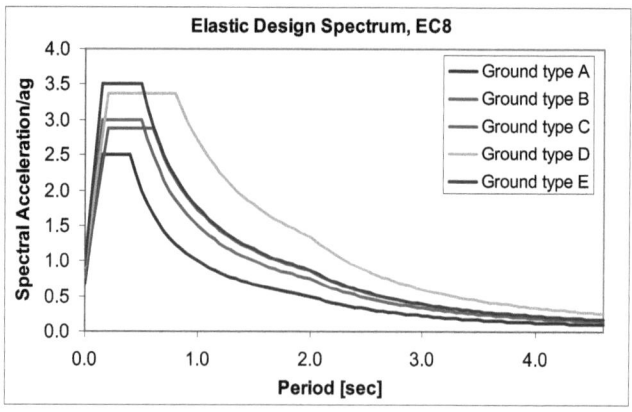

Figure 4-10: An example of elastic design spectrum from EN 1998-1 [27]

It is also important that hazard maps are prepared for both long periods and short periods, for instance 0.2 and 1.0 sec. This is not the case in many countries of the world and only long period hazard maps are available. These maps are not adequate for design of low-rise buildings.

4.2.3. Softening

When designing new buildings or retrofitting existing ones for seismic-resistance many architects and civil engineers think of strengthening buildings by increasing their lateral stiffness. A strengthening always stiffens the building, thereby raising the Eigen-frequencies. Under certain conditions however, it may prove more beneficial to soften a structure rather than to strengthen/stiffen it [8]. Therefore, it is preferred to soften buildings, while fulfilling load bearing capacity and serviceability of the building.

The same concept can be observed in case of structure-to-the-ground connections. Photos from earthquakes in harbors, where usually containers of goods are stored, show that, although these containers had been simply laid on the ground without any special connection, they are not damaged nor overturned. This is due to the low height of these containers in comparison with their length and width. There are just a very few photos available like figure 4-11, showing this fact. This is why there has been nothing catastrophic to be photographed in these cases. This is self a proof of the fact.

Figure 4-11:   Good behavior of containers in earthquake [29]

Base isolating systems utilize this concept in a more advanced way. By installing special soft seismic bearings above the foundation (base isolation), a frequency shift towards the lower area of the design response spectrum can be achieved (figure 4-10). As a result, and because damping is usually also increased, a significant reduction of the seismic induced forces and thereby the damage potential is achieved [8]. However, relative displacements increase in case of using base isolations, notably, which requires sufficient clearance around the isolated buildings. In addition, connection of pipelines between the building and the ground must be flexible enough to accommodate displacements without damage. These conservations have made the use of base isolating systems a challenging task for residential buildings, but attempts are observed in research [30]. However, the concept of base isolation is discussed for areas with a high level of seismicity in chapter 5.

4.2.4. Diagonal steel bracing

In design of buildings for seismic forces, the following systems are used to provide the lateral resistance of the structure:

- Braced frames
- Moment resisting frames
- Shear walls
- Mixed systems

For these systems, quality of connections plays a very important role, but this fact is usually underestimated. On the other hand, RC shear walls demand precise design, detailing and construction for a good performance, which involve big challenges in developing countries. In addition, RC shear walls are not appropriate for low-rise buildings, because they make the building very stiff and increase the seismic loads in the structure. Therefore use of RC shear walls is excluded for this work. However, among these systems, braced frames work with the simplest types of connections. In other words, braced frames, provide the lateral stiffness by simply hinged connections due to their geometry. Generally, braced frame systems lead to less amount of material. As mentioned before, and because of the good performance of these systems in earthquake, the solution presented in this work, supports strongly these systems for houses and schools. Braced systems can be divided further into concentric braced frames, in which the neutral axis of diagonal member meets those of column and beam in a point, and eccentric braced frames, in which there is an intentional distance between the diagonal-to-beam or diagonal-to-column connection and the beam-to-column connection, and is known as link in the literature. This type of braces has the ability of dissipating energy by plastic deformation and behaves more ductile. This can be very advantageous for the building in an earthquake. The concept of eccentricity is discussed in chapter 6.

In braced frames, diagonal members play the key role and have to be designed adequately. Since the seismic forces behave as a cycle of pushing and pulling forces, these elements get both compression and tension forces. Even those diagonals intended to work only in tension, has to remain functional under compression for the next cycles. The deformation can go into the plastic region of steel. This, combined with dynamic effects, must not lead to the failure of the structure. It is necessary to check compatibility between the deformations of the bracing and those of the other structural and non-structural elements. This can indicate the need for stiffer non-slender bracings or other bracing systems. Steel truss systems with eccentric connections and compact members behave much better than trusses with centre connections and slender members [8].

4.2.5. Ductility of structures

Shear walls of masonry and brick material have poor ductility. They are the main component of the structure destroyed first leading to the final collapse of the building (figure 4-12). In contrary, steel generally possesses good plastic deformation capacity or strain ductility. Nevertheless steel members and steel structures may show low ductility or even brittle behavior under cyclic actions, particularly due to local instabilities and failures. For instance column and beam elements with broad flanges may buckle in plastic zones or fail at poor welds. Therefore, certain requirements must be fulfilled appropriately and additional measures have to be considered during the conceptual design of the structure and the selection of the members' cross-sections [8] [21].

Ductile structures with large inelastic deformation capacity without failure offer substantial advantages in comparison to similar brittle structures. Ductile structures provide more safety with the same amount of material. Moreover, ductile structures can resist even earthquakes larger than the design earthquake. However, ductility requirements must have been considered in design, fabrication and erection. Currently, the capacity design method offers a simple and efficient approach to ductile structural design: The structure is «told» exactly where plastic hinges can be developed and where not. Hence, a favorable plastic mechanism is created. A large and predictable degree of protection against collapse can be achieved by good capacity design [8]. Ductility concept is explained more in chapter 6.

4.2.6. Floor and roof diaphragms

Floor and roof diaphragms are safety-relevant for the distribution of earthquake forces. Very steep roofs constrain the diaphragm action of roofs. Therefore the earthquake induced forces can not be conveyed equally to the horizontal load bearing system. Very heavy slabs have also been a cause of fatality in earthquakes. As it can be observed in figure 4-12, besides their high vulnerability, heavy diaphragms in combination with inadequate shear walls can cause the so called pancake collapse, which makes rescue operations after earthquakes extremely difficult.

When planning the floor or roof, all beams of that floor should be positioned in the same elevation. Application of the beams of a floor/roof level in different elevations interfere the functionality of the roof diaphragms and also divide the columns into short ones, as explained before. Hence, having rooms in different levels of a storey or half-stories should be avoided in buildings. Moreover, according to the performance based design rules, it is advised in codes that columns should not be weaker than beams.

Figure 4-12:   Pancake collapse of heavy slabs, Islamabad, Pakistan [AP]

4.2.7.   Connection of slabs

Connections in prefabricated buildings are often designed for gravity loads only. Such buildings can therefore be very vulnerable to earthquakes. Short support lengths, weak or missing dowels, and unsatisfactory overturning restraints of girders are frequently the cause of collapse. In case of prefabricated concrete floors, connections must be designed and constructed to guarantee a diaphragm action.

In addition, the connection of the floor/roof slab to the vertical elements must be in a way to distribute seismic forces during the earthquake properly without any collapse or rupture. As shown in figures 4-13 and 4-14, lack of appropriate seismic-resistant connections is a major cause of damage and death.

Figure 4-13:   Collapse of unfastened slabs [31]

Figure 4-14: Failure of a pre-cast concrete roof in the earthquake in Sichuan, China [34]

4.2.8. Infill walls

The poor performance of stiff and brittle infill walls during earthquakes has been addressed in many related literatures and is discussed by Peyvandi in [41]. According to [8], it is still a common opinion that filling in frame structures with masonry walls improves the behavior under horizontal loads including seismic actions. This is true only for small loads, and as long as the masonry remains largely intact. The combination of two very different and incompatible construction types performs poorly during earthquakes. The frame structure is relatively flexible and somewhat ductile, while unreinforced masonry is very stiff and fragile and may fully destroy under the effect of only small deformations. At the beginning of an earthquake the masonry carries most of the earthquake actions but as the shaking intensifies the masonry fails due to shear or sliding (friction is usually small due to the lack of vertical loads). The appearance of diagonal cracks is very characteristic of a seismic failure.

Two basic cases can be identified: either the columns are stronger than the masonry, or vice-versa. With stronger columns the masonry is completely destroyed and falls out. With weaker columns the masonry can damage and shear the columns, which often lead to collapse [8]. Generally, it is advised to decouple stiff masonry walls from the main structure. In some researches like [32], attempts have been made to prepare solutions for decoupling infill walls from the main structure. Figure 4-15, shows the main concept of the solution and a connection designed for this purpose, during an experiment done by Bauhaus University Weimar and Rudolstädter Systembau GmbH [32, 90]. In [8], it is suggested to create a joint between infill and the main structure by means of a very flexible and sound proof material. The joint thickness should be calculated for a specific building. Nevertheless, choice of the right material and construction of these joints are complicated and sensible, i.e. not applicable to developing countries. In this context, using local low price materials of low stiffness, for instance adobe infill walls with good ductility and damping characteristics, are advantageous, therefore preferred in this thesis. As observed in last earthquakes, in big wall panels, the low stiffness of walls lead to the out-of-plane collapse, which is highly live threatening (see figure 4-16). Many references, for instance [9], [31] and [33] have strongly called for reducing the risk of masonry and adobe infill walls. The risk increases with the size of the infill panel and the wall slenderness. In the contrary, as illustrated in figure 4-16, dividing the wall panel into small fields improves the performance of the wall during an earthquake and decreases the risk. Therefore, this concept is implemented in product development of houses and schools during the present work.

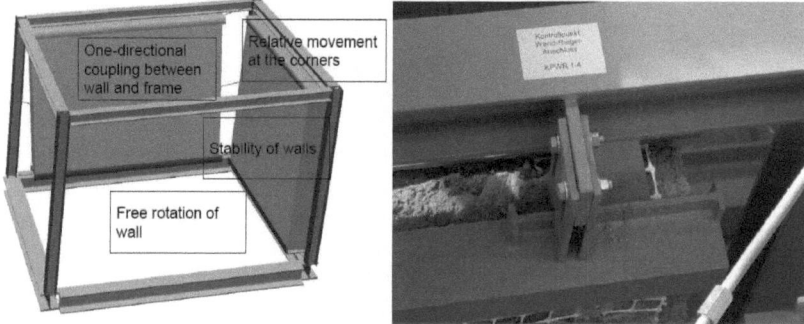

Figure 4-15: A connection designed to decouple the infill from the main structure [32]

Figure 4-16: Good performance of small fields of infill walls in comparison with large fields during the earthquake in Bam, Iran [31]

4.2.9. Foundations

According to the principles of the capacity design method, the foundations should be able to transfer the overstrength sectional forces of the plastic zones to the ground without yielding. Here, the foundation does not necessarily mean the fix anchoring of the structure to the ground, but to bear and transfer structural actions to the soil in a safe and stable way. As addressed in [8], foundation structures should always remain elastic since plastic deformations generally lead to an unpredictable behavior and additional displacements and stresses in the building structure. Besides, repairs are usually substantially more difficult to be executed in the foundations rather than in the building, itself (figure 4-17). The reinforcement must therefore be strengthened directly below the plastic zones and detailed accordingly.

Figure 4-17: Foundation failure of a slim building in earthquake in Adapazari, Turkey [12]

To ensure that seismic forces can be transferred to the soil, it is also advised to study the force path in the foundation structure. The allowable soil stresses under dynamic actions may be higher than the corresponding static stresses, but care should be taken to ensure that plastic deformations of the soil are avoided under all circumstances. This should be mentioned that plastic deformations of the soil may lead to the collapse of the whole structure. On top of this, the soil can change its specifications significantly during an earthquake going into liquid state. This makes overturning more probable. An example of foundation failure during earthquake is illustrated in figure 4-17.

4.3. Constructional aspects

4.3.1. On-site fabrication

Uncontrolled on-site fabrication makes even theoretically safe structures unsafe. On-site welding is a typical cause of this problem (figure 4-18). The quality of welding on site is usually much lower than in workshop. Reasons are:

- Transporting of the required technology to the construction site is not possible in terms of economics. The mobile technology does not end up with the same quality as in the workshop.

- The conditions can be controlled and adjusted for a good result in workshop. This is not possible on site for instance to exclude wind when welding with gas. Components can be positioned favorable for welding in the workshop. In the contrary, this flexibility does not exist on site, and components have to be welded in their actual positions, even overhead. For instance upward welding, which is often used on site, has a very low quality and is observed to be a major cause of failure.

- There is a lack of welding experts in housing sector in developing countries and the experienced welders are absorbed mostly by the heavy industry and for pipelines. There is also no authoritative supervision on site to control the welding quality. It is notable to mention that execution of welding needs expertise and its quality cannot be assured only by observation.

Consequently, welding on site should be absolutely forbidden; otherwise the vulnerability of the buildings will be high as shown in figure 4-18.

Figure 4-18: Poor quality of welding on site proved after earthquake in Bam, Iran [31]

4.3.2. Separation of adjacent buildings

Buildings should not transfer any force during vibration to their adjacent buildings. Buildings have to be designed separately and the unfavorable interaction between different adjacent buildings must be avoided. Based on the maximum displacement of the structure expected during the design earthquake, joints can separate the buildings. The result should be applied conservatively, since small joints cause the so called pounding effect, in which buildings will apply impact loads to each other. Hence, local and global failures can occur in the structure.

Implementation of joints is getting even more important for buildings not attached to the ground, for instance those having base isolation systems. In these cases, the building demands a free margin to be moved. For adoption of this concern for low-rise houses and schools, it is advised to construct buildings with enough distance from each other. This is not difficult in the small to middle towns and villages, since usually enough land is available.

4.3.3. Non-structural elements

Facade elements and those non-structural elements attached to the structure, experience the same amount of deformation as the main structure. If these components can not bear such deformations, they will collapse even with a relatively weak earthquake. Collapse and falling down of these elements can cause deaths. In addition, repair of damage in the main structure due to the collapse of non-structural elements is very costly. Facade elements should also be designed for seismic loads and the structure should give the possibility for connection of facade elements. Figure 4-19 is an example of damage caused in the facade of a house during the Bam earthquake.

Figure 4-19: Damage due to the failure of facade after earthquake in Bam, Iran [31]

5. Conceptual design of earthquake-safe houses and schools

At this stage the working principles are searched and evaluated to specify a principle solution. The subsequent steps for this phase include abstracting the essential problems, establishing working functions, searching for suitable working principles and then combining those principles into an overall working structure of the product [35]. A solution may have in turn different variants. All possible solutions will be gathered and evaluated to select the most promising ones. Brainstorming with experts from the field, talking with customers to know about their requests and learning from existing structures are helpful tools to get feedbacks from the market and to improve the product.

For evaluation during the conceptual phase, both the technical and economical characteristics should be considered as early as possible [35]. The costs can not be evaluated precisely due to the existence of unknowns in the design; but, the engineer should have a rough judgment of the prices when he searches for possible solutions. Furthermore, solutions are given a weight or a quantitative value, which makes the evaluation between different solutions possible. In other words, design starts with a focus on the most dominant evaluation criterion to formulate the base solution, which has to be modified later during judgment against other criteria. For case of earthquake-safe houses and schools, because earthquakes have a horizontal load effect on the building, this criterion can be the lateral load-bearing capacity of different structural elements. Characteristics of the wall panels and combination of these elements should be optimized in a way, that the structure gets lower seismic forces by behaving soft enough, while providing the necessary stiffness to reduce the displacements to the serviceability limit. Stiffness is not the only criterion for earthquake-resistant design, but this is used as the first one, before other criteria, e.g. ductility and performance based objectives, are implemented.

5.1. Structural concept

The high seismic performance of traditional half-timber houses used in Europe, especially in Germany, can inspire the modern technology today [38]. This type of building shown in figure 5-1 is known in Germany as "Fachwerkhaus" or truss house, which is translated as half-timber structures in English references. As it can be seen in the pictures, structural elements of these buildings are visible and are used as architectural elements. The reason is that these buildings are traditionally made of timber elements, due to availability of wood in these regions. However, the high quality and safety of these buildings does not owe to their material, but their geometrical concept. Dividing the wall elements into small fields, with many diagonal elements, brace the whole structure against horizontal loads with high redundancy and also confines the infill walls. The latter increases the load bearing capacity of infill walls, since this way, the out-of-plane behavior of the wall is highly improved. This improvement is very important for adobe or masonry walls, with low resistance to the out-of-plane collapse. However, because wood resources are limited in many countries, wood can be substituted with steel, which is produced and recycled all over the world, including many developing countries. This requires in turn substituting the term half-timer structures with the more general term, "systematically braced frames", in this dissertation.

Figure 5-1: Typical "Half-timber" structures in Germany and Europe [39]

Due to the unique architecture of these buildings, infill walls are divided into small panels. As mentioned in chapter 4, this aspect is very advantageous for resistance against seismic loads. Furthermore, having many diagonal members, these houses can bear horizontal loads of an earthquake. For such reasons and also because of its simplicity, this type of building is prepared as a pattern for houses and schools for developing countries. The result will be a steel structure, which is different from conventional steel frames in several cases. The load bearing system consists of load bearing walls, which are created from connection of smaller wall panels together. Because the distance between columns is reduced intentionally in this type of structures (i.e. to 1.5 m), even smaller profiles can be used. In the case of low-rise buildings, these profiles can be even light-gauge cold-formed steel profiles, which are light on the one hand, and have a high strength-to-weight ratio on the other hand. The latter is achieved as a result of the cold-forming process of the steel, which increases its yield strength at corners. Because of their good attributes for prefabrication, similar wall panels of family houses made of cold-formed steel profiles get increasingly their market. In addition, cold-formed profiles are more cost-efficient for developing countries, because their cost is less affected by market fluctuations. They are all built by steel construction enterprise from steel strips and sheets, which can be bought and stored in large batch sizes and formed into different shapes.

Currently available systems of wall panels, also called steel shear walls, in the literature, are discussed in many references, e.g. [37], [84] and [85]. These systems generally include vertical studs connected in small spans to the lower and upper tracks, which have the beam function. Some diagonal elements or strips brace the whole wall. In order to restrain elements against buckling, the whole wall is covered with for instance gypsum or fiber-cement plates. An example of such elements is illustrated in figure 5-2.

Figure 5-2: A wall panel made of cold-formed steel profiles [18, 37]

In buildings built with such wall elements, because the whole wall element is prefabricated in desired length, the building arrangement should be predefined. In addition, the total weight of the wall will be high and delivery and installation have to be done by crane. Such shortcomings make these systems unsuitable for construction and assembling by local labor. Instead, a wall panel, created from assembling of smaller frames, has many advantages. Small frames can be carried and assembled by field workers. However, connections must be designed and executed in a way to guarantee the integrity of the whole wall element. Different types of possible connections are discussed in [84].

In order to provide the rigidity of slabs, floors and roofs are intended to be concrete plates. Following the rule in this work that all safety relevant components have to be manufactured in the workshop, floor/roof elements are also intended to be prefabricated plates. These elements will be connected together appropriately to form a rigid diaphragm.

Buildings are further decided to be low-rise of up to three stories maximum. As mentioned in chapter 4, higher buildings are costly not only due to the higher construction techniques required, but also because of special design demands they have. For instance, in most of the countries, buildings of four stories and higher have to be equipped with lifts and separate escape rout.

The solution supported by this work, includes manufacturing of the main structural-relevant elements, including wall panels and floor/roof components, with high technology in the workshop and under strict supervisions. Products will be labeled or tagged afterwards to be distinguished from other products with uncertain quality. Being safety relevant, these products are then delivered to the construction site for assembling done by well trained personnel of the manufacturer. However, implementation of some technical measures can assure the sound assembly of components by giving the proving possibility at any time. After assembling of structural components, infill walls and finishing works are accomplished by local labor. This reduces the costs and increases the cultural acceptance of the product, by giving the local hand workers the chance, to provide the building with a favorable appearance. This leads to individual products made from combination of a specific number of modular products.

5.2. Structural safety-relevant components

5.2.1. Walls

As mentioned before, wall systems of the systematically braced frames are intended to be an assembly of smaller prefabricated panels, which can be handled by two persons on the site. The whole wall including smaller panels must behave as a single element with integrity in different loading situations. The lateral resistance of a wall is provided by panels, braced with diagonal members. As shown in figure 5-3, it is practically not possible and not necessary to have diagonal members in all panels due to the demand for openings, like doors and windows. Therefore, within a constant width, three types of panels shown in figure 5-4 are required:

- Frame braced with a diagonal member
- Frames with horizontal door bar
- Frames with horizontal window bars

According to the simplification rule, the variation should be limited as much as the functionality is not reduced. Consequently, a single constant width of 1.5 m was assumed primarily for all wall panels. Panels of other width can be manufactured by order, but the cost will be more, because they are not a standard module. However, these modules cannot be combined together in an arbitrary way in sense of stiffness and serviceability. Each wall system should have enough stiffness and displacement capacity, intended in the design. Lower values of stiffness may lead to large displacements under horizontal loads, mainly earthquakes, while higher values increase the resulting actions in members, unintentionally. In addition, fulfillment of ductility concerns is very important for seismic-safety of the building. The underlying principles, namely stiffness and ductility requirements, are explained because of their importance in the following sub-sections for systematically braced frames. The calculation of stiffness is the same as in other types of structures adopted here for wall panels, while the ductility requirement is developed within this work.

5. Conceptual design of earthquake-safe houses and schools

Figure 5-3: Front view of a family house with different types of panels *[Courtesy: N. Nasserian]*

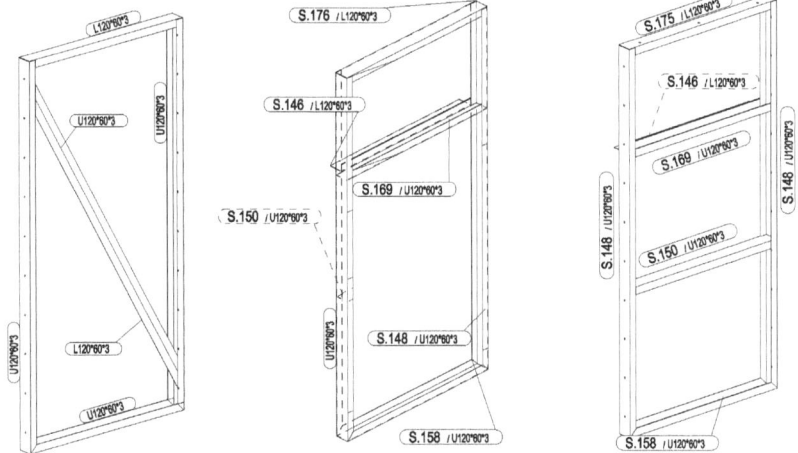

Figure 5-4: Three types of wall panels: braced panel, door panel, and window panel

Furthermore, in assembling of panels together, these rules must be considered, too:

- Combination of wall panels should comply with the regularity laws. This means that the center of stiffness of the building in the plan should be equal to its center of gravity, in order to avoid torsional effects in the building.

- In each assembly of walls at least two braced frames must be considered, in opposite directions. This way, at least one bracing will work always in tension and the other in compression. Although braces are desired to work in tension, they must not fail nor experience plastification upto target displacement of the storey in opposite load direction, defined in codes [27].
- Braced panels should be distributed over the whole length of a longer wall. Implementation of bracings in only one region increases the vulnerability of the structure in case of fire, because the failure of that region, removes the stability of the whole structure.

From a geometrical point of view, two types of cross-sections are considered for wall panels with the assets and drawbacks mentioned here.

Panels made of L-profiles

Considering the fact that even unskilled people should be able to construct infill walls on site, barriers have to be minimized before. As observed experimentally in case of sample houses during summer schools, these barriers include matching of bricks with steel structure in spite of tolerances of both materials, especially bricks. Using an open-formed profile like L-profiles can decrease the construction problem to a great deal. Figure 5-5 illustrates combination of panels made of L-profiles with two types of infill materials, namely brick and adobe. As it is shown here, the flange of L-profile will be in the middle of the wall to confine the masonry or adobe infill and to prevent early collapse of the steel element in case of fire, as well as to minimize the thermal bridge.

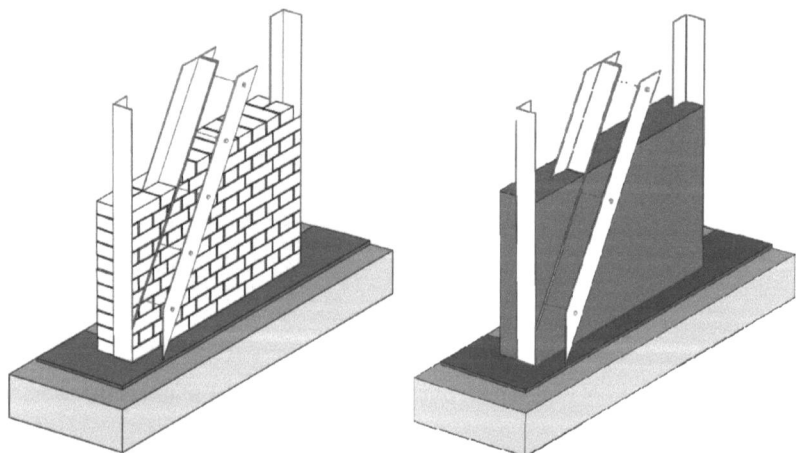

Figure 5-5: Wall panels made of L-profiles covered with decorative elements, filled with brick (left) and with adobe (right) *[Courtesy: A. Eghdam]*

However, from a structural point of view, L-profiles have a low bending and compression capacity. Moreover, due to the relatively high slenderness of components of an L-profile, they are exposed to local buckling. To overcome these drawbacks, cross-sections have to be larger in design, which lead to an increase of material with non-standard cross-sections and a higher cost. If they are used in combination with bricks, in a way that bricks provide elements with lateral support (confined masonry walls), this problem is solved. Hence, use of L-profiles is recommended if infill walls contribute to the load-bearing system. In this case, the load-bearing-system will be a reinforced masonry wall, in which steel panel helps confining the infill and the wall prevent L-profiles from local buckling. This interactive effect between wall panel and infill, leads to a system with a capacity higher than the capacities of its components. This interaction between panel and infill is discussed by Peyvandi in [41].

Panels made of L-profiles and U-profiles

In comparison to the L-profiles, U-profiles are composed less to local and distortional buckling. This is due to the fact that the web of a U-profile is supported at both ends, while in a L-profile both legs are supported only at one end and will buckle with a smaller moment and/or axial force.

In addition, proper back-to-back connection of U-profiles forms an I-section, which is very suitable for bending. Therefore, the total amount of steel required will be reduced and the final price will be lower. However, the complexity of matching bricks into the open side of the profile should be solved by increasing the height of web in the section up to a few millimeters to cover the tolerances of bricks (figure 5-6). From a constructional point of view, it is also advantageous for bricking of the most upper row to make horizontal elements from L-profiles.

Figure 5-6: Wall panels made of U-profiles covered with decorative elements and filled with brick

5.2.2. Floors and roofs

Floor and roof slabs are structural-relevant components of a structure, which contribute in good performance of a building under seismic loads. This means that according to the concept of this work, they have to be manufactured with a guaranteed quality and delivered to the site. Especially in developing countries, construction of these elements at the site leads to serious deficiencies in the seismic resistant of the building. Besides quality and robustness, the shorter construction time is also very advantageous for houses and schools. Therefore, the volume of construction works has to be reduced by minimizing the amount of form working on the site.

In traditional construction of floors, the process has to be interrupted at each floor for hardening of the concrete, which increases the construction time. A major improvement can be achieved if concrete elements are manufactured in the workshop under desired conditions and delivered to the site for assembling. The size of these elements depends to the level of rigidity and integrity of the final floor or roof as well as to the total weight of each element. While having a less number of elements is in favor of integrity of the whole slab, for delivery and construction processes it is better to have a lighter weight, which means a higher number of elements. Considering that wall panels are manufactured with a constant width of 1.5 m, the integrity demand will be fulfilled if floor and roof elements have the same width and transfer of actions to the wall panels is assured by appropriate connections. However, the total weight has to be reduced by either using light-weight concrete or by optimizing the cross section of the element, which means to eliminate that part of concrete that has no significant influence on the load bearing capacity of the plates.

Light-weight concrete, known also as Autoclaved Aerated Concrete (AAC Technology), is produced by implementing very small air bubbles in concrete during casting. This technology leads to light-weight concrete with a unique cellular structure that provides superior energy efficiency, fire resistance and acoustical properties as well as lower seismic induced forces due to the light weight. AAC was developed by Dr. Johan Eriksson in 1923 at the Royal Technical Institute in Stockholm, Sweden. In 1945, Josef Hebel invented a method to produce ACC by incorporating steel into the production process. The self weight of ACC including reinforcement is only 6.7 kN/m^3, while that of the normal concrete is about 25.0 kN/m^3 [40, 46].

Disadvantages of light-weight concrete are its lower strength and the complexity of its manufacturing in comparison to the normal concrete. Therefore, this type of elements is generally used for residential buildings and not for public buildings like schools with higher loads. Figure 5-7 shows construction with light-weight concrete plates of Ytong Company [40]. The vertical load bearing capacity of these elements is presented in table 5-1.

Concrete plates optimized in shape are also known as hollow concrete plates, because the volume of concrete is reduced by implementing continuous holes without having significant decrease of the bending capacity of the plate. Holes can be implemented by putting pieces of polystyrene shaped in the form of holes in the concrete prior to casting. This way, not only the weight is reduced, but also the holes will not place any waste materials during construction.

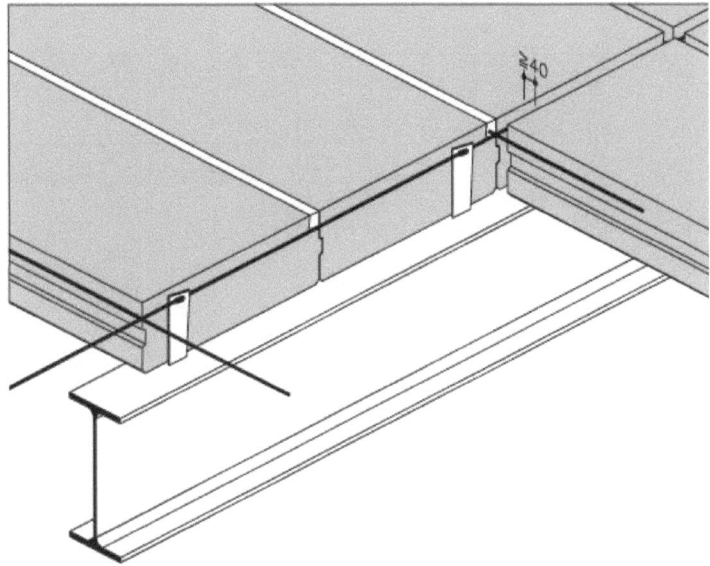

Figure 5-7: construction details of light-weight concrete plates [40]

Porenbeton-Deckenplatten, Feuerwiderstandsklasse F 30 nach Zulassung Z-2.1-4..1									
Zulässige Stützweiten in Abhängigkeit von der vorhandenen Belastung die ()-Werte gelten für Ausführungen in Feuerwiderstandsklasse F 90									
Plattendicke d	Nutzlast[1] g_2+p [kN/m²] (ohne Eigenlast der Deckenplatten)								Eigenlast g_1
	2,50	2,75	3,00	3,25	3,50	3,75	4,00	4,25	4,50	
[mm]	maximale Stützweite [m]									[kN/m²]
200	5,25 (4,33)	5,17 (4,26)	5,10 (4,20)	5,03 (4,15)	4,96 (4,09)	4,89 (4,04)	4,78 (3,99)	4,67 (3,94)	4,57 (3,88)	1,34
240	6,06 (5,13)	5,97 (5,06)	5,89 (4,99)	5,82 (4,93)	5,75 (4,87)	5,68 (4,81)	5,61 (4,76)	5,50 (4,70)	5,44 (4,65)	1,68

1) Ständige Last aus Bodenaufbau g_2 = 1,0 kN/m².

Table 5-1: Load bearing capacity and span for light-weight concrete panels offered by Ytong [46]

Similar products are currently in use for construction of public buildings like schools, office buildings and multi-story car parks. Their appropriate implementation can increase the efficiency as mentioned by Goldbeck in [45]. Adding reinforcement is also possible in hollow concrete plates with either pre-stressed bars or by normal bars. Although pre-stressed plates have a higher capacity, their manufacturing and construction is more complex. Further, pre-tension force has to be controlled and maintained during the life cycle. Therefore, it is preferred to use normally reinforced hollow concrete plates to end up with more economic buildings for developing countries. Figure 5-8 shows a typical hollow concrete plate developed during this work to be used in schools.

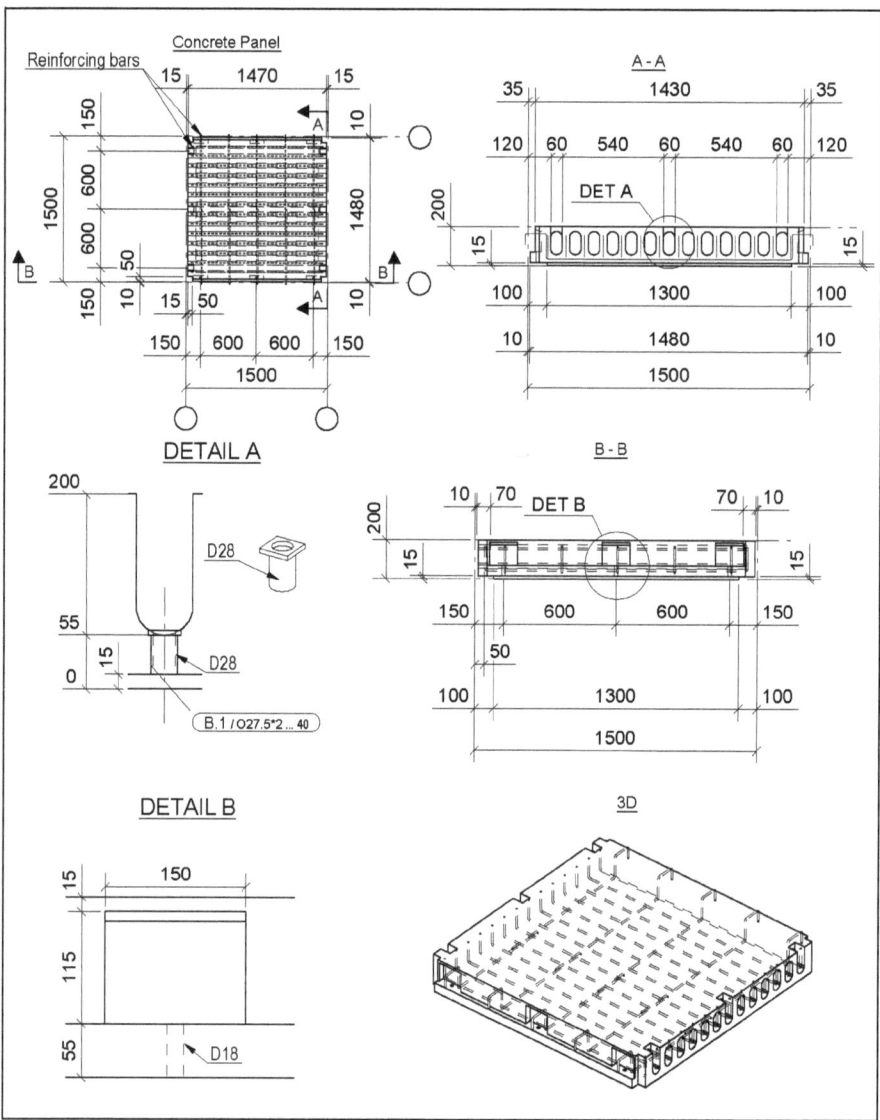

Figure 5-8: Hollow concrete plate suggested in this work

To get an understanding of the vertical load bearing capacity of currently produced hollow concrete plates, table 5-2 shows such data for Echo pre-stressed elements with different spans [49], while the capacity of normal reinforced elements is investigated in [41].

To apply prefabricated products in earthquake prone regions of the world, modifications are required to provide lateral resistance of these elements. This includes firstly the rigidity of the whole slab under seismic actions. A rigid diaphragm distributes lateral forces of an earthquake between vertical elements proportional to their stiffness alone, which is desired in the design. There is no perfect rigid diaphragm, but the limit under which a diaphragm can be considered as rigid defined in codes is achievable. EN 1998-1 defines for instance where a diaphragm can be considered as rigid. Modeled with its actual in-plane flexibility, its horizontal displacements must nowhere exceed those resulting from the rigid diaphragm assumption by more than 10% of the corresponding absolute horizontal displacements in the seismic design situation [27]. Diaphragms made by assembly of separate prefabricated plates should be designed carefully for this aspect, since they tend to behave separately if this issue is not considered precisely. In this regard, continuous joints between adjacent plates should be designed to transfer shear as well as compression and tension, resulted from shell and membrane action of the slab.

	Without structural topping					
Dead load kN/m²	1,5 kN/m²					
Live load kN/m²	1,5	2	4	5	7	10
120 mm	485	475	440	430	400	350
150 mm	655	640	595	580	545	490
200 mm	885	870	830	785	745	680
240 mm	1065	1045	980	950	900	850

Table 5-2: Maximum span for different live loads in pre-tensioned hollow concrete plates prepared by Echo [49]

According to EN1992-1 [22], shear transfer in connections may be achieved in different ways. Three main types of connections shown in figure 5-9 are as below:

a) Concreted or grouted connections

b) Welded or bolted connections

c) Cast-in-place reinforced topping of the plates

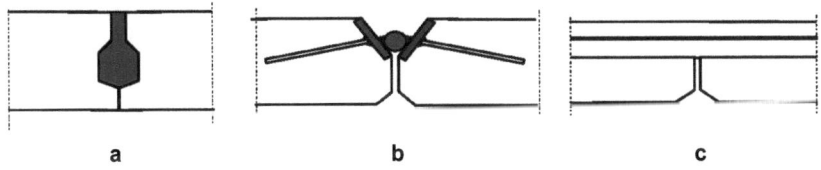

Figure 5-9: Connections transferring shear between concrete elements [22]

Quality of both cast-in-place concrete and welding cannot be assured on site. On the other hand, quality of bolted connections depends to the quality of the bolt itself and to its proper construction, both an open question for many regions in developing countries. Concreted connections in contrast, can be used with a higher certainty, since the functionality will be achieved in a simpler way. However, to assure the transfer of all actions through joints in concreted connections, the joint should be designed precisely and requirements have to be implemented in the prefabricated part. These actions include shear in different directions as well as compression and tension. DIN 1045-1 [50] suggests the form shown in figure 5-10. The gap has to be filled with a concrete having at least the same compressive strength as the plates. In order to avoid any mistake, appropriate concrete mixture can be delivered by manufacturer to the site in dry packs. Being a safety-relevant task the filling of the gap is done by skilled personnel of the industrial manufacturer of the structure. Therefore, it is part of the contractual duty including responsibility.

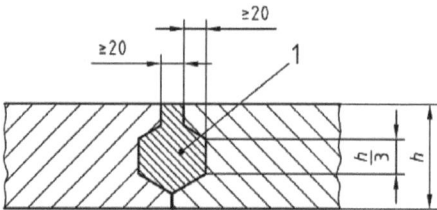

Figure 5-10: Form of joint recommended in DIN 1045-1 to transfer shear actions [50]

The tension due to the bending in the slab should be resisted by reinforcement along the joints as shown in figure 5-11. Because of the varying nature of seismic loads, bending can happen differently over different cross sections, which means that reinforcement has to be calculated for the maximum load and implemented similar along all joints. Joists and beams used to connect the plates together can be used simultaneously to carry this tension load, if the shear transfer is provided along these elements.

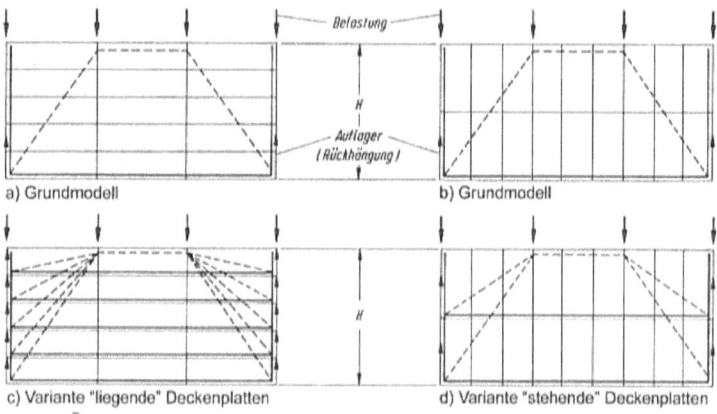

Figure 5-11: Equivalent load-bearing system in floor diaphragms [51]

5. Conceptual design of earthquake-safe houses and schools

The distance between centerlines of the beams is taken equal to those of wall panels (1.5 m) and the size of concrete plate is adjusted to this distance (see figure 5-8), considering the connection requirement. Figure 5-12 shows hollow concrete plates connected to the girders.

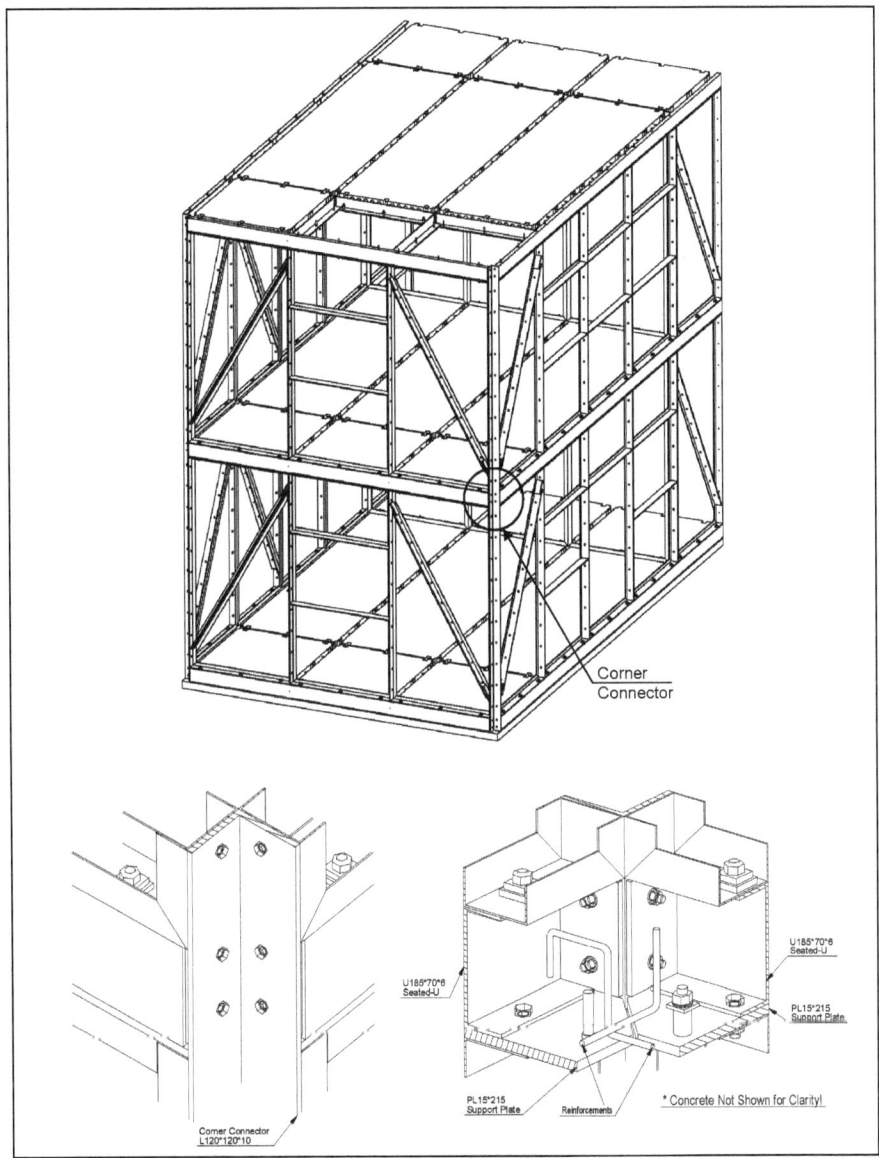

Figure 5-12: Combination of hollow concrete plates to the steel structure and detail of corner connectors

The construction method used here is the platform method. However, corner connectors made of L-profiles can be continuous along the height of the building as shown in figure 5-12, since there is no barrier, which makes interruption necessary. This should not be confused with balloon method which requires continuous columns.

Considering that concrete plates have to be easily put on joists, these elements should create no barrier against plates. Therefore the upper flange has to be smaller than the lower flange or completely removed, to make the distance between plates and the volume of gap smaller. The roof construction will distribute loads depending to the length-to-width ratio of a slab. The capacity of recommended hollow concrete panels is calculated in [41]. However, steel T-girders have the function of bearing loads during construction of floor/roof assembly. A T-profile made of plates welded together is shown in figure 5-13. In order to make use of temporary supports before the joints are filled unnecessary, the maximum length of these girders with S235 steel plates up to 15 mm thick and a total section-height of 150 mm, can be 4.5 m. This result is based on ultimate state calculation and a maximum deflection limit of L/300, applying the weight of plate and a live load of 3 kN/m² for schools. The thickness of web can be even less for girders of 1.5 m and 3.0 m length.

The T-profiles intended for beams can be provided also by cutting I-profiles, e.g. from IPE series. The steel construction enterprise can even substitute them based on static calculations at different times due to the market situation. The developed method makes the use of temporary supports and form working for abovementioned spans unnecessary on site, which is in favor of an economic and fast construction. Construction of larger spans demands appropriate set up of props.

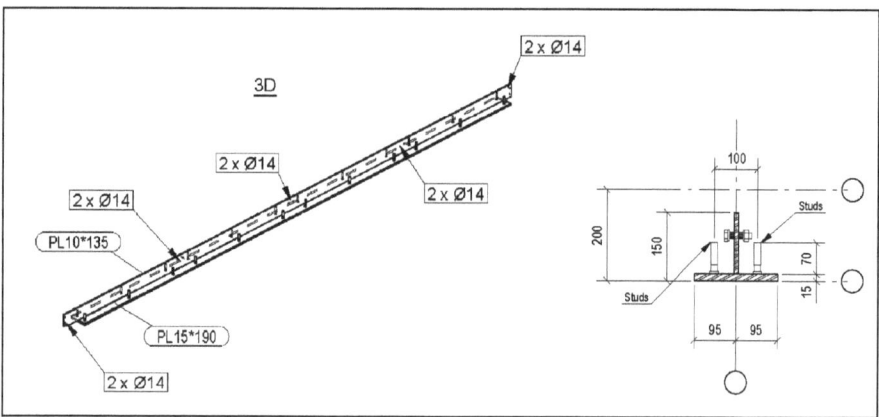

Figure 5-13: Suggested form of beams and joists

Openings and special corners

For different reasons openings are required in the floor and roof slabs. While large openings can be implemented by eliminating a roof element, for smaller ones extra (special) elements can be used. Generally, small openings are used for passing pipes and cables through the slab. To keep this flexibility while working with developed system, extra modules with small openings are shown in table 5-3. This way pipes or cables can pass through the slab wherever required with only minor modifications in their path.

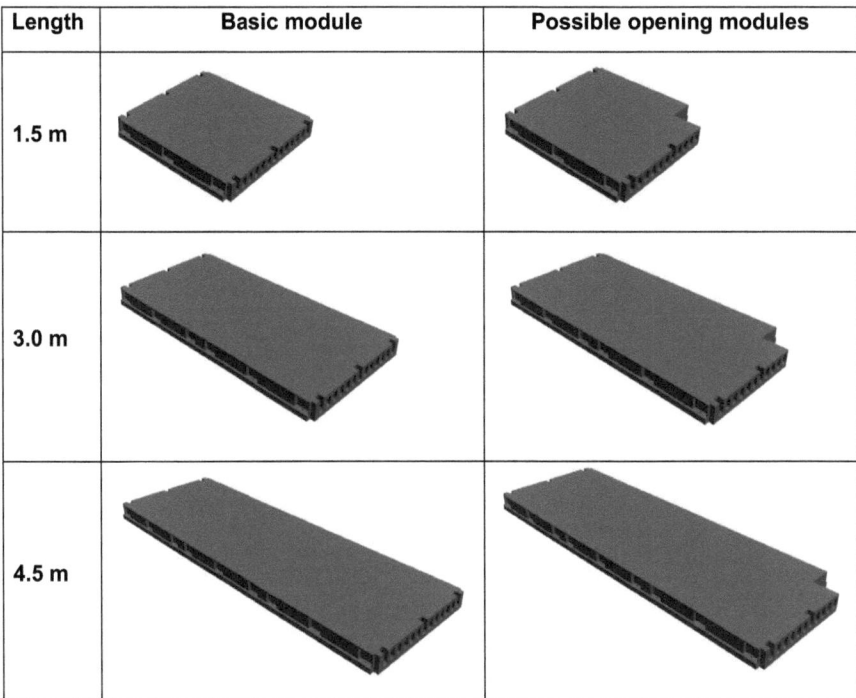

Length	Basic module	Possible opening modules
1.5 m		
3.0 m		
4.5 m		

Table 5-3: Basic modules and modules for small openings with different lengths for floors and roofs

Assembly

Hollow concrete plates have to be adjusted in their place on the beams using small cranes, installed for instance on delivering trucks, to reduce total costs. As shown in figure 5-14, the studs welded on the flange of the joist or beam to transfer shear can be used at the same time for adjustment, by fitting into the holes of the plate. These holes are created in the concrete plates by embedding steel tubes shown in detail A of figure 5-8. Threaded studs are welded automatically and precisely to the beams by stud welding machines. All studs used in this work comply with characteristics provided by Köco Stud Welding [52]. After adjustment, threaded studs are fastened using two layers of locking washers (e.g. Nordlock® fastening security), which prevent bolt connections from loosing [53]. Procurement of these components seems not to be a serious problem even in developing countries.

Product development of earthquake-safe houses and schools

To be able to use the same concrete elements everywhere, supporting element has to be modified in different positions. Different variations shown in figure 5-14 include middle T-girders in all 1.5 m, seated-T members used on inner frames, as well as seated-U used on outer frames.

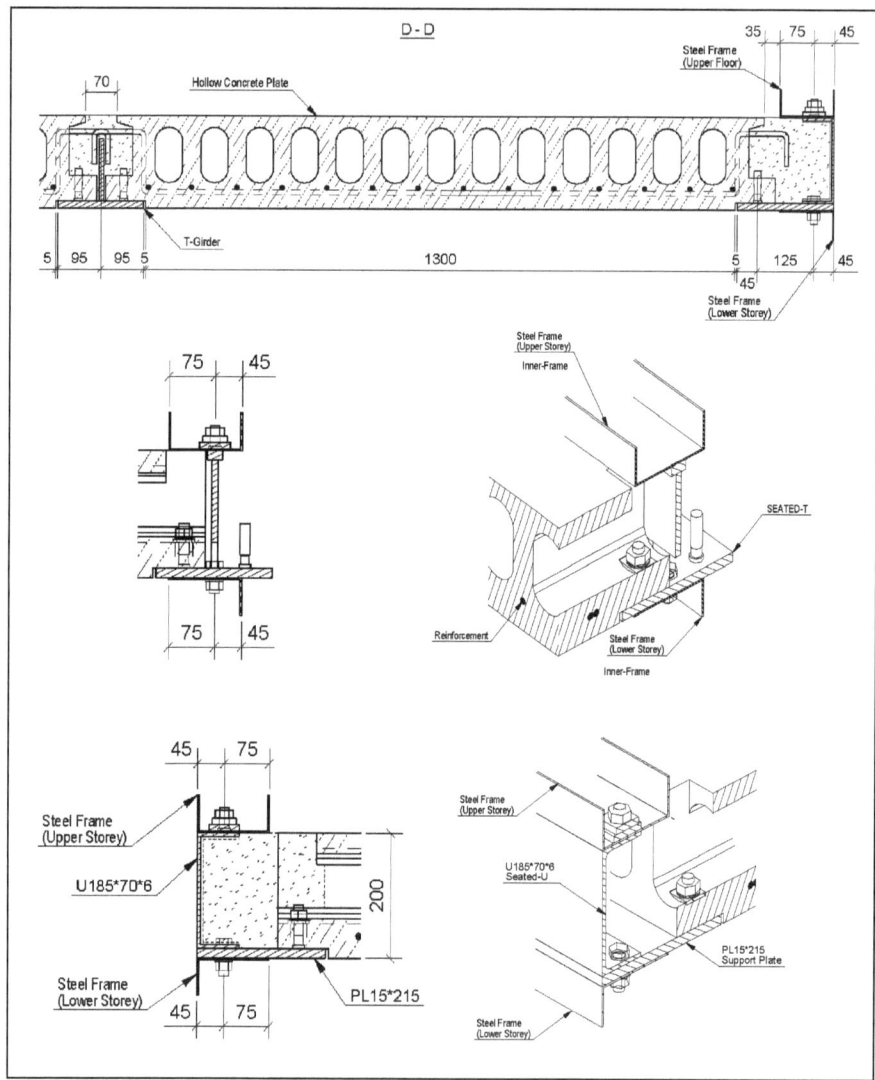

Figure 5-14: Assembly of roof on middle T-girders, seated-T and seated U-profiles

5.2.3. Base isolation

In order to prevent or reduce the seismic induced loads in the structure, seismic isolation systems are developed detaching the building from the ground. Since the seismic isolation is mostly installed at the base level of the building, it is known as base isolation. In earthquake engineering, level of hazard is different in each region and this has to be considered in product development. In other words, products (houses and schools) have to be designed according to the demand in the region of applicability. However, to minimize the variation in products for sake of economic, buildings can be designed for a certain level of hazard, where they can be built without any change, while in other areas with higher level of hazard the effect on the building is reduced for example by using base isolations.

In addition, even in relatively moderate earthquakes in areas with poor housing, many people are killed by the collapse of brittle, heavy, unreinforced masonry or poorly constructed concrete buildings. Modern structural control technologies such as active control or energy dissipation devices can do little to alleviate this, but it is possible that seismic isolation could be adopted to improve the seismic resistance of poor housing and other buildings such as schools and hospitals in developing countries [30].

The principle of seismic isolation is to introduce flexibility at the base of a structure in the horizontal plane, while at the same time introducing damping elements to restrict the amplitude of the motion caused by the earthquake. The concept of seismic isolation became more feasible with the successful development of mechanical energy dissipators and elastomers with high damping properties. Seismic isolation can significantly reduce both floor accelerations and interstory drift, which are two major indicators of damage to the buildings. In addition, they provide a viable economic solution to the difficult problem of reducing non-structural earthquake damage. There are three basic elements in any practical seismic isolation system. These are as follows [70]:

- A flexible mounting so that the period of vibration of the total system is lengthened sufficiently to reduce the force response
- A damper or energy dissipator so that the relative deflections between building and ground can be controlled to a practical design level
- A means of providing rigidity under low (service) load levels, such as wind and minor earthquakes

Seismic isolation achieves a reduction in earthquake forces by lengthening the period of vibration in which the structure responds to the earthquake motions. The most significant benefits obtained from isolation are thus in structures for which the fundamental period of the building without isolation is short, less than one second. Therefore, seismic isolation is most applicable for low-rise and medium-rise buildings and becomes less effective for high-rise structures [70].

The theory of seismic isolation [72] shows that the reduction of seismic loading by an isolation system depends primarily on the ratio of the isolation period to the fixed-base period.

Since the fixed-base period of a masonry block or brick building may be around 1/10 second, an isolation period of 1 second or longer would significantly reduce the seismic loads on the building and would not require a large isolation displacement. For example, the current UBC code for seismic isolation (ICBO 1997) has a formula for minimum isolator displacement which, for a 1.5 second system, would be around 15 cm [30].

A large number of base isolations are developed over the time and patents are still registered every year for new products. However, the most major types of seismic isolations and their design considerations are discussed by Naeim and Kelly in [72]. DIN EN 1337-3 [28] contains the requirements and rules regarding different types of supports to be used in European countries. The World Housing Encyclopedia reports on base isolations [9], contains application of the following two different types of isolations in housing, which look feasible also for schools and their implementation can be extended there:

- Sliding-based Isolation systems
- Rubber-based isolation systems

Sliding-based Isolation systems

These systems include layers of materials with a very low coefficient of friction, which can slide over each other if the excitation exceeds a certain level. The function is similar to the sliding clutch in a car. Because of the importance and high influence of friction, these systems are known also as friction-based isolations (figure 5-15). The most commonly used materials for sliding bearings are unfilled or filled polytetrafluoroethylene (PTFE, or Teflon) on stainless steel, and the frictional characteristics of these systems are dependent on temperature, velocity of interface motion, degree of wear, and cleanliness of the surface [72].

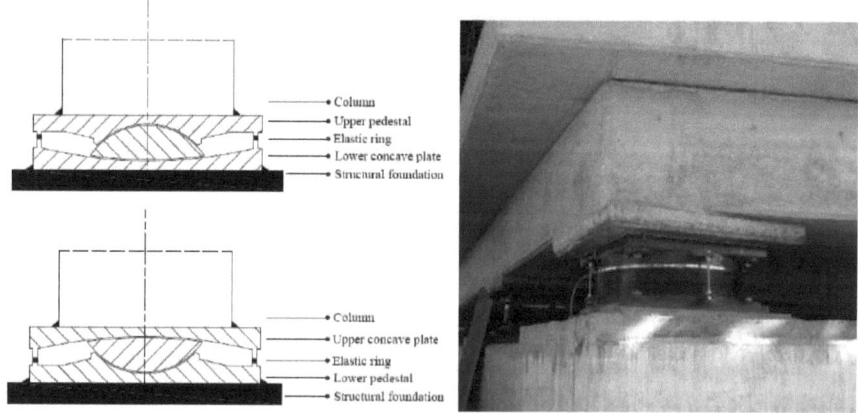

Figure 5-15: Friction-Pendulum System isolators (downward and upward) [73, 75]

A numerical simulation for Friction-Pendulum Systems (FPS), shown in figure 5-15, is developed and discussed by Jamali in [73]. Using a concavity radius of infinity, the application and implementation of these systems are tested afterwards by Rezaido [74] for housing in developing countries, in which all walls are decoupled from the ground using a layer of steel strip laid over a Teflon strip. A schematic detail of the system is presented in figure 5-16. As evident in the photo, building can move simply over the ground surface. Therefore, buildings have to be constructed with enough distance from each other, which seems to be no problem in villages or small towns. A similar case of implementation is reported in [9] as sliding-belt isolation system, used in Kyrgyzstan, with the difference of having horizontal and vertical restraints for displacements larger than a certain limit.

Figure 5-16: Application for housing in developing countries [74]

Rubber-based isolation system

Conventional rubber bearings consist of many steel shims embedded in rubber (figure 5-17). The rubber is vulcanized and bonded to the steel in a single operation under heat and pressure in a mold. The essential characteristic of the elastomeric isolator is the very large ratio of the vertical stiffness relative to the horizontal stiffness. Therefore, the steel shims prevent bulging of the rubber and provide a high vertical stiffness but have no effect on the horizontal stiffness, which is controlled by the low shear modulus of the elastomer. The manufacturing process of these elastomers is well known, besides they are easy to model, and their mechanical response is unaffected by rate, temperature, history, or aging [72].

The adoption of base isolations to developing countries is explicitly studied among others by Kelly [30]. As mentioned there, the problem relating to developing countries is that conventional isolators are large, expensive, and heavy. An individual isolator can weigh one ton or more, which require cranes for installation. The primary weight in an isolator is that of the steel reinforcing plates used to provide the vertical stiffness of the rubber-steel composite element. A typical rubber isolator has two large end-plates around 25 mm thick and 20 thin reinforcing plates around 3 mm thick. The process of preparing and embedding plates in rubber is a complex and time consuming one, reflected in the high price of these isolators.

 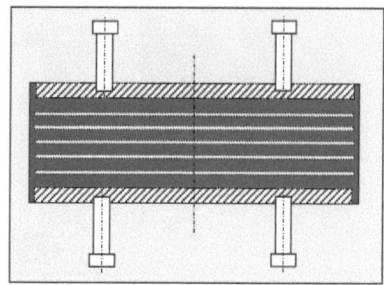

Figure 5-17:   Conventional rubber bearing [72, 75]

To extend this earthquake-resistant strategy to housing and public buildings, the cost and weight of the isolator must be reduced. Both weight and cost can be reduced, if steel plates are substituted with cheap fiber reinforcement, while having the same vertical stiffness. Therefore, attempts are made to substitute steel with fiber reinforcement (figure 5-18). The fiber-reinforced isolator is significantly lighter and can be made by a much less labor-intensive manufacturing process in an industrial workshop. The advantage of the strip isolator is that it can be easily used in buildings with masonry and adobe walls. Furthermore, using fiber reinforcement instead of steel plates makes it possible to build isolators in long rectangular strips, with individual isolators cut to the required size. All isolators are currently manufactured as either circular or square. Rectangular isolators in the form of long strips would have distinct advantages over square or circular isolators when applied to buildings where the lateral-resisting system is made of walls.

When isolation is applied to buildings with structural walls, additional wall beams are needed to carry the wall from isolator to isolator. For this reason, a concrete slab is assumed at the base level of the building (see figure 5-15), to assure that all displacements are the same at this level, even if isolators do not tend to.

Figure 5-18: Testing of rubber bearings developed for developing countries [30]

5.3. Cost analysis and calculation

5.3.1. Cost factors

Early identification and estimation of cost factors during product development helps to end up with economic designs. Total costs can be effectively controlled during conceptual and structural design phases, while the only small influences are possible after the solution is defined.

The overall cost of a product can be divided into direct costs, which are directly confronted when producing a component, e.g. cost of materials and labor forces, and indirect costs, which may be appeared later, e.g. running costs. Furthermore, costs can be divided into variable and fixed costs. Variable costs are those depending to the batch size or ordering amount. Fixed costs are incurred in a certain period of time, e.g. management salaries, rent of the place and interest on borrowing. Designers can directly influence variable costs by appropriate choice of working principles, materials, production times, batch sizes, production processes and assembly methods. Therefore, for making decisions during the design process, variable costs (direct and indirect) are of interest [35]. As mentioned in [35], direct variable costs have to be estimated by designer, even roughly, during design process, while indirect variable and fixed costs can be estimated later, by multiplication of some factors by direct variable costs.

5.3.2. Fundamentals of cost calculation

The variable part of the manufacturing cost (VMfC) is the basis of decision making. It comprises direct material costs (DMtC) and production labor costs (PLC), including assembly costs. Therefore, the variable manufacturing cost can be written as:

$$VMfC = DMtC + \sum PLC$$

Direct material costs are determined by either weight (W) or volume (V) of materials and corresponding specific cost (c), which represents the costs per unit weight or volume.

$$DMtC = c_w.W = c_v.V$$

On the other hand, production labor costs (PLC) can be calculated considering the time required for each individual production process or assembly operation, multiplied by the labor cost factor (c_L). Production time consists of a primary time (t_p), a secondary time (t_s) and a set-up time (t_{su}), as well as distribution and recovery time. Generally, the last two parts of time are taken into account by a constant factor on the basic time (t_b), which is the sum of primary and secondary times and results in a time per unit.

Therefore, for the calculation of costs, primary, secondary and set-up times are important. Using the labor cost factor (c_L), the following simplified equation can be used for a particular production operation:

$$PLC = c_L(t_p + t_s + t_{su})$$

For cost estimation based on quick and overall calculations, strict determination of direct costs, according to their individual dependencies, is not necessary and can even take too much effort. Basically, detailed calculations at this stage can not and are not required to be very precise, because there are still unknown parameters in the development of the product. Based on [35], three basic methods of cost estimation are explained below.

Comparing with relative costs

In this method prices and costs are related to a reference value. Considering the fact that absolute prices change due to market fluctuations and inflations especially in developing countries, from time to time, results obtained by this method are more generic and have a long-term validity in comparison to the absolute costs. Therefore, application of this method is recommended in [35] for cost calculation.

Relative material costs (c^*) are usually compared with structural steel S235 and can be calculated from the following equation, in which (c_w^*) or (c_v^*) are specific material costs corresponding to the weight and volume, respectively:

$$c_{w,v}^* = \frac{c_{w,v}}{c_{w,v}(\text{reference value})}$$

A cost calculation based on this method is presented in tables 5-4 and 5-5 for main components of houses and schools. The costs used in these tables are official prices published yearly as "Fee Structure for Buildings and Structures", used for public contracts in Iran in Rials, for the Iranian calendar year from March 2008 to March 2009 [82]. These data change from year to year and are used here only for sake of comparison without guarantee.

5. Conceptual design of earthquake-safe houses and schools

The reference value used here for the time period mentioned is taken equal to 9,351 Rials and is calculated as the average price of I-profiles of height 120 to 300 mm, published for the first nine months of the corresponding year by government of Iran [83]. The reason for this decision is that these profiles are produced usually in the country and their prices show fewer changes in comparison to other sections usually imported. The variation in price per weight of different I-profiles is shown in figure 5-19 for the time from March 2008 to November 2008.

Type of component	Type of element	Total area [m²]	Cost [Rials/m²]	Total cost [Rials]	Relative costs
Hollow concrete plate of 20 cm thickness and 1.5 m length	Concrete Steel rods	2.25	80,600 133,164	480,969	51.4
Hollow concrete plate of 20 cm thickness and 3.0 m length	Concrete Steel rods	4.5	80,600 133,164	961,938	102.9
Hollow concrete plate of 20 cm thickness and 4.5 m length	Concrete Steel rods	6.75	80,600 133,164	1,442,907	154.3

Table 5-4: cost calculation for hollow concrete plates based on relative costs

Type of component	Type of element	Total weight [kg]	Cost [Rials/kg]	Total cost [Rials]	Relative costs
Braced panel with L profiles	Columns	66.12	11,900	1,777,758	190.1
	Beams	29.58	10,700		
	Bracings	59.16	11,400		
Door panel with L profiles	Columns	66.12	11,900	1,261,587	134.9
	Beams	44.37	10,700		
Window panel with L profiles	Columns	66.12	11,900	1,419,840	151.8
	Beams	59.16	10,700		
Braced panel with U and L profiles	Columns	22.62	11,900	679,110	72.6
	Beams	12.72	10,700		
	Bracing	24.02	11,400		
Door panel with U and L profiles	Columns	22.62	11,900	473,334	50.6
	Beams	19.08	10,700		
Window panel with U and L profiles	Columns	22.62	11,900	541,386	57.9
	Beams	25.44	10,700		
T-girder of 1.5 m length	Beam	49.46	10,700	529,222	56.6
T-girder of 3.0 m length	Beam	98.91	10,700	1,058,337	113.2
T-girder of 4.5 m length	Beam	172.21	10,700	1,842,647	197.1

Table 5-5: cost calculation for different steel components based on relative costs

The costs presented here do not consider the method of manufacturing. For an automated industrial production, the enterprise has to finance the costs required for the production line. However, these costs are one-time costs, and have to be compensated later by mass production of component and constant contracts the steel manufacturing enterprise signs. This way the company makes benefit by using the economy of scale, in which the advantage increases by the increasing the production. In this regard, looking for constant investors demanding for similar type of buildings, e.g. school constructing organizations and building societies, has a decisive importance.

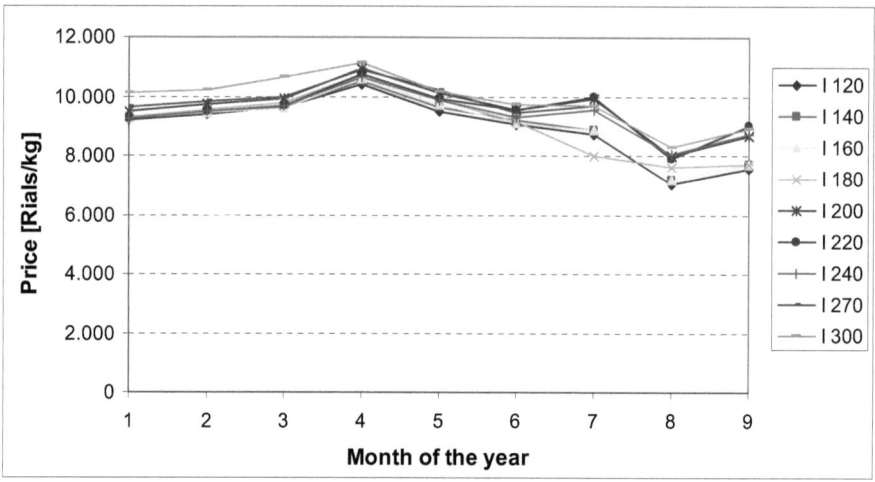

Figure 5-19: Variation in price of different I-profiles from March 2008 to November 2008

Estimating using share of material costs

In this method, manufacturing cost (*MfC*) is related to the material cost (*MtC*) with applying a coefficient (*m*). The general form is therefore:

$$MfC = MtC/m$$

For application of this method, only one material has to be in each analysis and the manufacturing process should be simple and easy structured. Therefore, for family houses and schools with several manufacturing, construction and assembly processes, application of this method will not present realistic results.

Estimating using regression analysis

In this method, the relationship between costs and characteristic values are defined by use of statistical analysis. For this reason, all influencing parameters are gathered and implemented in a power series with parametric coefficients and powers. Using statistical data processing methods on the existing data, these coefficients and powers will be defined. To get better answers, enough data should be available and the resulted relationship is valid only for similar products manufactured under the same conditions. Because of these limitations, extrapolation will normally lead to unacceptable results. These draw backs make this method unsuitable for product development of earthquake-safe houses and schools.

5.4. Non-structural components

5.4.1. Cultural acceptance and attractiveness

Attractive appearance is one of the most important decision arguments of costumers and a characteristic of successful products. This is the first issue that all types of customers care about. While aesthetic has to be considered for all types of products in daily life of humans, it has obviously a special meaning for houses. On the other hand, people living in different regions have different tastes and aesthetic must be defined according to local interests. Therefore, design has to be done by local architects and artists in order to make houses and schools not only acceptable, but attractive to people. Architectural design by local handworkers creates simultaneously job opportunities. The issue of acceptability and attractiveness of a new building type in a region depends to the social behavior and culture and has to be tackled with a wide approach. First, the building has to be designed in a way that it looks very attractive to the people and second, it should be attempted to introduce the building to the habitants with all its features and the advantages in comparison with other types of houses. Furthermore, construction of schools can be used as good teaching examples on how to build also family houses earthquake resistant.

To make the appearance of the house or school attractive to people while keeping structural elements visible, the structural elements can be beautified with decorative strips. Figure 5-20 shows decorative strips installed on the panels filled on the left side with brick and on the right side with adobe in the sample building constructed at Bergische University Wuppertal. The pattern is intended to follow the Iranian art and was drawn and colored by students attending the DAAD summer school 2008.

To bring the cultural acceptance for houses and schools, following techniques can be used:

- Publication and distribution of brochures containing attractive images from houses and schools to provide customers and decision makers with the final appearance of the product and the contribution of the product in urban planning. Figure 5-21 shows some example images which can be used in this regard, prepared during the present work. Therefore, the catalogue containing all possible combinations makes decision easier for customers.

- To show houses built and furnished in real scale can be a decisive tool, since the potential customers can get a feeling of a house when they visit it. As a more advanced idea, a show house in order to increase the public awareness on seismic risk can be launched in a public place in cities of developing countries. The building of this learning center itself is a systematically braced frame, which is discussed before in [14].

Figure 5-20: Attractiveness by Iranian pattern to decorate the building for a better appearance

- Architectural design competitions can involve local architects from early stages and the manufacturing enterprise organizing the competition can supply the modules in order to comply with attractive designs.
- Model buildings in small scales or carton houses showing the new concept can be used as educational material for students. This makes not only students but also families familiar with appropriate and safe buildings. Figure 5-22 shows a carton model of a German half-timber building used for this purpose.

5. Conceptual design of earthquake-safe houses and schools

Figure 5-21: The final appearance of houses in urban and rural areas *[Courtesy: N. Nasserian]*

- In this regard, media, in its different forms, can play a very important role. Media can remind people the earthquake hazard and inform them on how to build houses and schools earthquake-resistant. Installation of posters in public places, video teasers showed for instance as TV advertisements, and articles in local magazines or newspapers are some possible ways.

Figure 5-22: Carton house model of a German half-timber building

6. Structural design of earthquake-safe houses and schools

During this phase the conceptual design will be worked out toward realization and will be converted to layouts. Pahl and Beitz use the term embody design for this phase in [35] and mention that the very main elements of embodiment design are clarity, simplicity and safety. Layouts designed based on these three principles have to be evaluated against technical and economic criteria and the financial viability. Through a series of methods explained in standards or developed during the project, each layout has to be checked for optimum functionality, durability, production, assembly, operation and costs. Because of the diversity of the design criteria at this stage, embody design is also known as the phase of "design for x". These criteria may already exist at the beginning of the project or can be detected in a later phase, when a modification on the product is aimed. The most promising layouts passing these criteria will be selected at the end of this phase for detail design. In other words, the part "design for x" aims at supporting the basic elements and helping design engineers to meet the specific requirements and constraints. This chapter starts with structural design of systematically braced frames, including seismic design, and continues with other dominant issues regarding houses and schools.

6.1. Structural features of the main components

6.1.1. Eccentrically braced frame module

The aim of this part is to investigate the quality of load bearing in systematically braced frames with eccentricity. While a properly designed and constructed moment resisting steel frame can behave in a very ductile manner, they are very flexible and the resulting displacement is usually large under lateral loads. On the other hand, concentrically braced frames have a large lateral stiffness, but their energy dissipation capacity is affected by buckling of the brace. However, eccentrically braced frames can be designed to have the required ductility and stiffness at the same time [36]. Eccentrically braced frames are extensively researched previously for conventional steel structures, and results are published among others in [36, 47 and 48]. The focus of these works is mostly high-rise buildings or industrial structures, in which the behavior of the small element created by the eccentricity between bracing and beam-column-connection (link) is not necessarily as in low-rise buildings. For instance, because the distance between columns are larger in high-rise buildings, it is possible to implement the link in the beam, while the bracing-beam angle is not too sharp. Doing the same in case of wall panels with small spans leads to a sharp angle between diagonal bracing and column, which decreases its efficiency. However, the methodology of these researches can be used for low-rise buildings as well. Figure 6-1 shows braced frames with eccentricity in beam and in column. For an economic fabrication, the cross-section of bracings and columns are selected to be the same.

According to [47], the characteristic feature of eccentrically bracing systems is that the axial forces induced in the braces are transmitted either to a column or another brace, largely through shear and bending in a segment of floor beam or the link. An eccentrically braced frame system is a hybrid, deriving its stiffness from the truss action and its ductility by inelastic deformation of the link. It is mentioned in [36], that the critical beam segment, or link, is designated by its length, e, which should be measured from the face of or the edge of a column. Links in eccentrically braced frames act as structural fuses to dissipate the earthquake induced energy in a building in a stable manner to prevent buckling of the braces. To serve its intended purpose, a link needs to be properly detailed to have adequate strength and stable energy dissipation. All the other structural components (beam or column segments outside of the link, braces, columns, and connections) are dimensioned following capacity design provisions to remain essentially elastic during the design earthquake [36]. For seismic applications, braces are designed such that not to buckle under extreme loading conditions. This basic requirement can be accurately estimated, and an eccentrically braced frame is dimensioned in a way that under severe loadings the major inelastic activity takes place in the link [47].

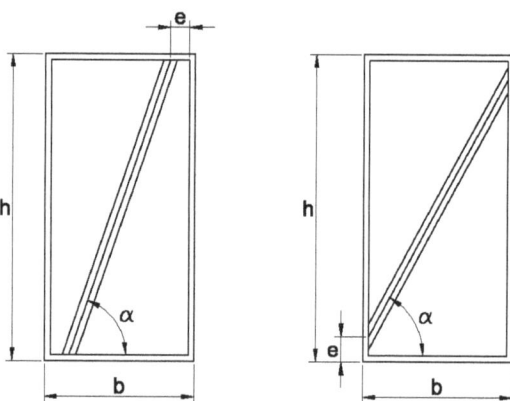

Figure 6-1: Eccentricity in beam and column in systematically braced frames

Elastic Stiffness

The variation of the lateral elastic stiffness with respect to the ratio of eccentricity (e) for a wall panel with a width equal to 1.5 m is shown in figure 6-2, for both cases of eccentricity in the beam and in the column. Here, the effect of buckling under compression is excluded, and the results are presented only for tension. As shown in this picture, under this assumption, the difference between two curves is negligible.

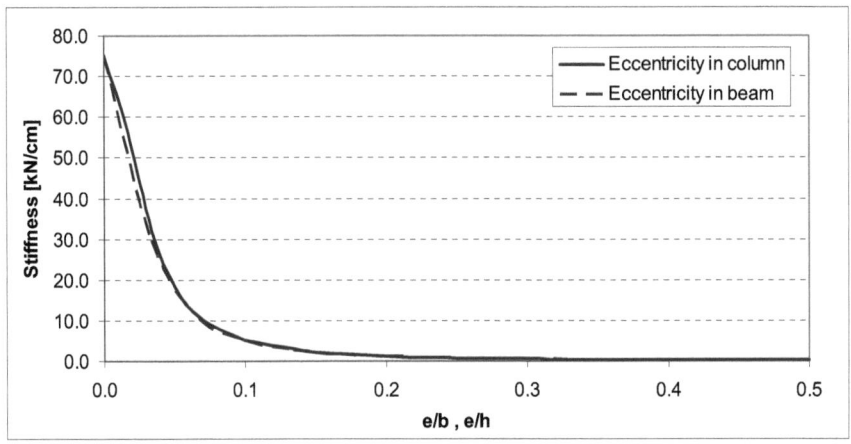

Figure 6-2: Variations of lateral stiffness with respect to link length in beam and in column

Figure 6-3 shows the degradation of stiffness with increase in inclination of diagonal member. As it is evident in this figure, the effective values will be achieved if angle α is between 50 to 75 degrees. The peak value in this graph corresponds to a concentric braced frame.

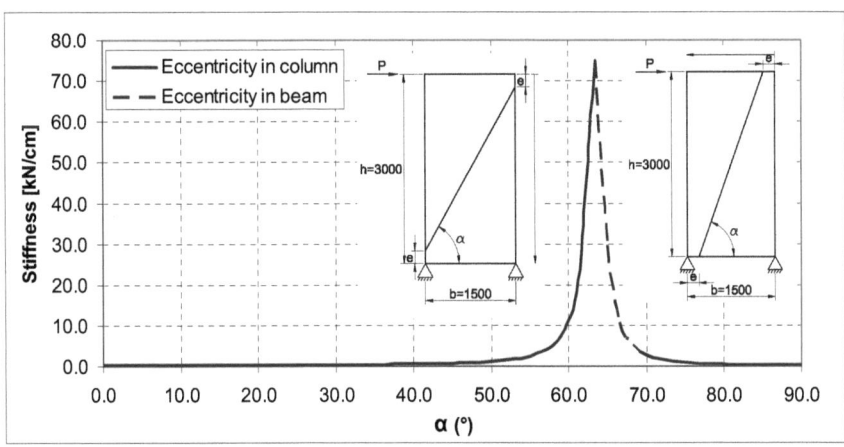

Figure 6-3: Variations of lateral stiffness with respect to angle α

Based on extensive studies carried by Popov, a categorization of the link lengths and their behavior into three regions is presented in [36], as a function of shear and flexural capacity of the link, V_p and M_p, as shown in figure 6-4.

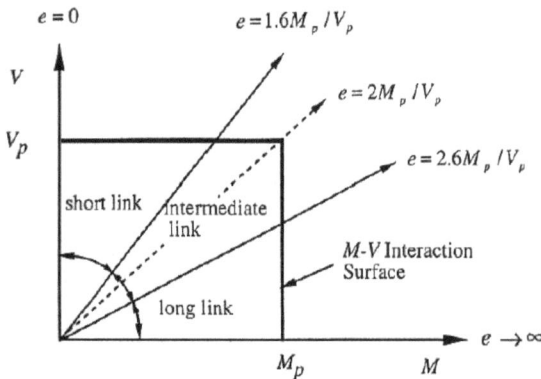

Figure 6-4: Classification of links [36]

Effect of concrete slab

One of the main requirements in eccentrically braced frames is that the link has the possibility to be deformed and to dissipate energy while deforming. In conventional steel skeleton structures, this behavior can be achieved if the eccentricity is implemented in beams. It is even recommended by performance based seismic design rules to consider eccentricity in beams rather than in columns, to prevent early yielding of columns due to large deformations. This fact is not true in case of systematically braced frames for two reasons. First, concrete slabs of floors are constructed between columns (platform construction) and are connected to the beams and links along their length. Therefore, links cannot deform easily and if it happens, the slab and consequently the whole system will be damaged under cyclic loads. Second, the high degree of redundancy in systematically braced frames causes forces in members to be much lower than the ultimate resistance limit of members. Moreover, redundancy helps the structure to remain stable even if a few members yield. Due to these reasons it is no problem in case of systematically braced frames to put eccentricity in columns rather than in beams. Therefore, this has been considered as an assumption for the solution developed through this work.

6.1.2. Ductility

Ductility is defined as the ability of the structure to exceed the yielding point and plastify without failure. Depending to the purpose of use, ductility includes several meanings, which are listed in table 6-1. For seismic design of structures, ductility is one of the most important aspects for safety and has to be addressed accordingly [42, 43]. This means that buildings have to be designed not only for strength, stability and serviceability criteria, but also to have enough ductility. The required level of ductility depends to the importance level of the structure and to the expected level of earthquake, for which the building is going to be designed.

6. Structural design of earthquake-safe houses and schools

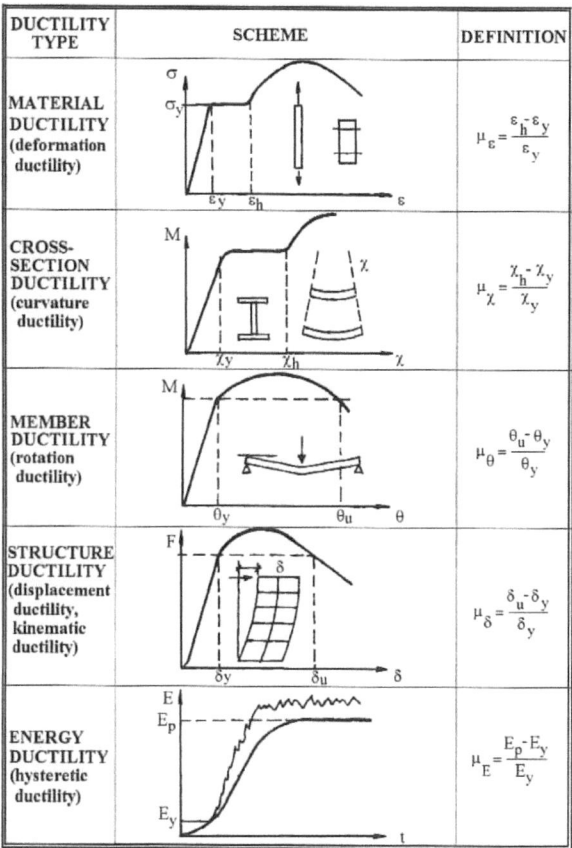

Table 6-1: Different types of ductility [42]

On the other hand, to make the design economic, codes take the ductility concept into account whenever dealing with seismic loads. Because of the relatively low probability of earthquakes, structures are allowed to be designed taking their plastic capacity into account. Due to the complexity of plastic design, a simplification is done in code in which effect of seismic load is reduced in proportion to the ductility of structure by the so called behavior factor. However, the reduction is done in a way to guarantee that the structure will not exceed its critical damage level even if the seismic loads exceed the elastic capacity of the structure.

As observed in systematically braced frames, structural elements are distributed in the whole structure with intentionally small spans. This leads to smaller cross-sections because of a smoother distribution of forces. However, construction aspects require that reduction is applied to the thickness of cross-sections rather than to their height or width. For example, whenever walls are assumed to be filled with adobe or masonry materials, the height of L- or U-profile can not be less than a certain value to provide enough space for bricking. On the other hand, reduction in the thickness of material, compose cross sections to different types of instabilities, namely, local, distortional and global buckling [71]. Degradation in resistance due to these types of buckling has to be considered in the design, while neglecting leads to underestimated results and will increase vulnerability.

In context of EN 1993-1 [21], cross sections with high probability of local and distortional buckling are categorized as class 4 sections. When buckling occurs, some part of the cross-section area will be buckle and its remaining part, which will bear the load, has to be recalculated. Reduction in the area causes reduction in the stiffness and demands new analysis of the structure under current load level. This means that for systematically braced frames made of thin-gauged profiles cross-sectional ductility differs due to the fact that it is based on the current level of load and remaining area of the section at that level. To calculate the remained part or effective area of the section, EN 1993-1-5 suggests implementation of a reduction factor ρ as follow [25]:

- For internal compression elements:

$\rho = 1$ \qquad for $\overline{\lambda}_p \leq 0.673$

$\rho = \dfrac{\overline{\lambda}_p - 0.055(3+\psi)}{\overline{\lambda}_p^2} \leq 1.0$ \qquad for $\overline{\lambda}_p > 0.673$, where $(3+\psi) \geq 0$

- For outstand compression elements:

$\rho = 1$ \qquad for $\overline{\lambda}_p \leq 0.748$

$\rho = \dfrac{\overline{\lambda}_p - 0.188}{\overline{\lambda}_p^2} \leq 1.0$ \qquad for $\overline{\lambda}_p > 0.748$

where $\overline{\lambda}_p = \sqrt{\dfrac{f_y}{\sigma_{cr}}} = \dfrac{\overline{b}/t}{28.4\varepsilon\sqrt{k_\sigma}}$

ψ is the stress ratio determined in accordance with EN 1993-1-5 tables 4-1 and 4-2 [25].

$\overline{b}$ is the appropriate width to be taken as follows:

b_w : for webs;

b : for internal flange elements (except rectangular hollow sections);

$b - 3t$: for flanges of rectangular hollow sections;

c : for outstand flanges;

h: for equal-leg angles;

h: for unequal-leg angles;

k_σ is the buckling factor corresponding to the stress ratio ψ and boundary conditions. For long plates k_σ is given in Table 4.1 or Table 4.2 of EN 1993-1-5 as appropriate;

t is the thickness;

σ_{cr} is the elastic critical plate buckling stress.

As nature of the problem requires, the load-deformation behavior has to be modeled up to the failure of the section. The prerequisite is therefore, a realistic model of the material, in which stress and strain capacity are considered up to the failure. For steel structures, different idealized models of materials are presented in EN 1993-1-5 [25] for different purposes of calculation. However, as mentioned in [87] none of these models are suitable for ductility purpose, because they all consider deformation capacity up to a limit lower than failure strain. To overcome this shortcoming, some idealized material models suitable for ductility estimations, are mentioned and investigated in [42]. The model used during this work is taken as that developed by Boeraeve et al in 1993, considering its high accuracy and its independency to laboratory experiments. This model idealizes the stress-strain curve introducing 5 points and uses a straight line representation for hardening range. This model is compared in figure 6-5 with experimental behavior of structural steel ASTM A36, which is similar to S235 of EN 1993. It can be observed that the idealized model complies with real behavior up to a certain limit, while neglects the capacity of the material after this limit. Excluding this part of capacity leads to an underestimation of ductility and is therefore conservative.

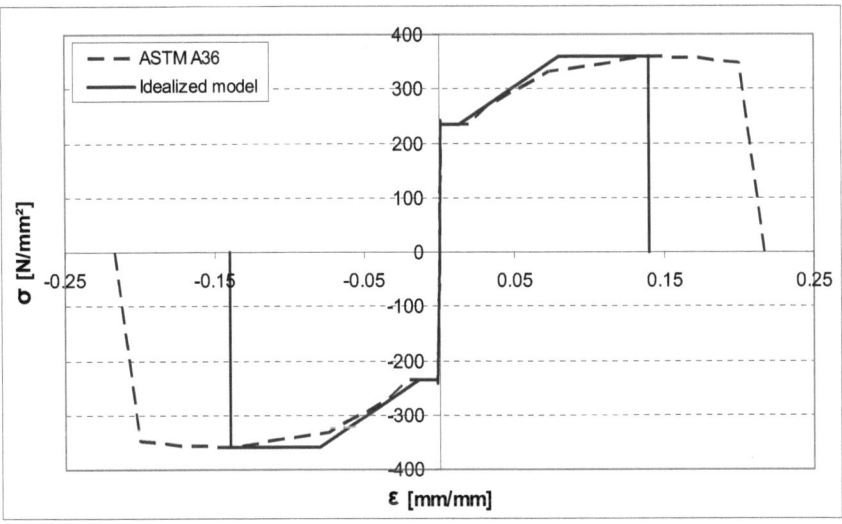

Figure 6-5: Idealized material model [42] vs. real stress-strain curve

Another assumption used here is that straight planes in the beam remain straight after deformation (Euler-Bernoulli beam theory). Therefore, stress distribution over height of the section under pure bending remains proportional to stress-strain curve. Depending to the location of neutral axis in the remained part of the section, credible distributions over the height are presented in table 6-2 with corresponding boundary condition.

	Boundary Conditions	a $0 \leq f_t < f_y$	b $f_t = f_y$	c $f_y < f_t < f_u$	d $f_t = f_u$
1	$f_c = f_y$ End of elastic range				
2	$f_c = f_y$ End of plastic plateau				
3	$f_c = f_u$ End of strain-hardening				
4	$f_c = f_u$ Collapse				

Table 6-2: Possible stress distributions over the height of section as a result of bending

In addition to the material, reduction factor should be extended for ranges other than elastic range. For that reason, the critical plastic buckling stress of the plate has to be calculated. In [88] critical plastic buckling stress is estimated by introducing a reduction factor η into the elastic buckling relationship. Consequently, the general relationship for buckling of plates will be:

$$\sigma_{cr} = \eta \frac{k_\sigma \pi^2 E t^2}{12(1-v^2)b^2}$$

where η is plasticity reduction factor, which can be calculated in accordance with table 6-3;

E is the modulus of elasticity equal to 210,000 N/mm² for structural steels;

v is the Poisson's ratio equal to 0.3.

Structure	η
Long flange, one unloaded edge simply supported	$\dfrac{E_{sec}}{E}$
Long flange, one unloaded edge clamped	$\dfrac{E_{sec}}{E}\left(0.428 + 0.572\sqrt{\dfrac{1}{4} + \dfrac{3}{4}\dfrac{E_{tan}}{E_{sec}}}\right)$
Long plates, both unloaded edges simply supported	$\dfrac{E_{sec}}{E}\left(\dfrac{1}{2} + \dfrac{1}{2}\sqrt{\dfrac{1}{4} + \dfrac{3}{4}\dfrac{E_{tan}}{E_{sec}}}\right)$
Long plate, both unloaded edges clamped	$\dfrac{E_{sec}}{E}\left(0.352 + 0.648\sqrt{\dfrac{1}{4} + \dfrac{3}{4}\dfrac{E_{tan}}{E_{sec}}}\right)$

Table 6-3: Values of plasticity reduction factors η for different boundary conditions [88]

Implementation of the abovementioned method is illustrated in figure 6-6, in which the moment-curvature graph is shown for a sample U-profile and an L-profile in pure bending. Due to the asymmetry of L-profiles, graph is concluded for both cases of the horizontal leg (flange) in tension and horizontal leg (flange) in compression. It should be mentioned that in case of L-profile, when flange is in tension, section fails very soon, and the graph drops to the zero. Therefore, the graph is very small in compare to the case of flange in compression.

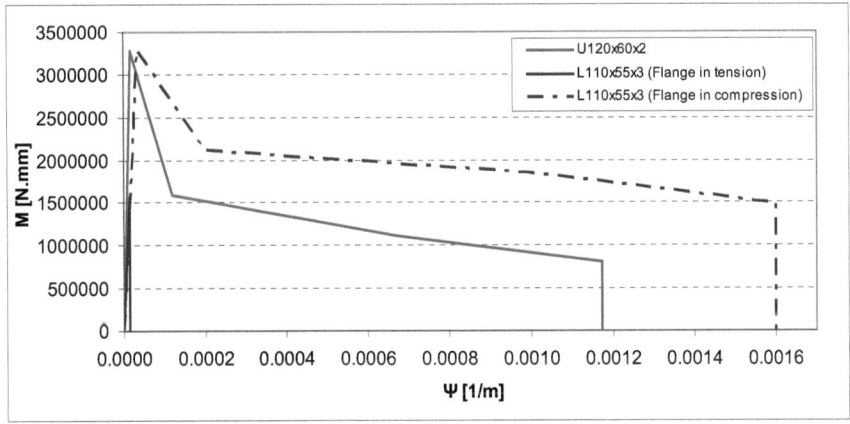

Figure 6-6: Comparison of the moment-curvature curve of a sample U-profile and a L-profile

6.2. Other important aspects

6.2.1. Design for production and assembly

Design for production means designing for the minimization of production costs and times while maintaining the required quality of the product [35]. The term production usually refers to:

- The production of components
- Assembly, including transportation of components
- Quality control
- Material logistics
- Operations planning

Because of the common and traditional method of costing by weights, designers usually think that the total costs would be reduced if weight is reduced. Considering manufacturing and assembly aspects, this assumption is often wrong. In addition, especially in developing countries, prices change rapidly because of variation in market influencing factors, mainly inflation. Engineers have usually less information about the market prices during design. For instance, it is observed that there is a tendency among structural engineers to select small cross-sections and use stiffeners in critical locations. Consequently, production and assembly prices including cutting, sand blasting and welding will be increased, although the total weight is less. Therefore, in such cases using larger sections will be economically more effective. Hence, engineers have to consider technical and economical aspects of different variants in design.

In addition, designers are usually not aware of the decisive influence they have on the costs and quality of assemblies. Attempts should be made to decrease the number of different tools required during assembling. A high number of different tools increases the costs and requires more time in the field. Whenever possible, several assembling and construction processes should be possible parallel at the same time. Processes designed in a line, in which a process has to wait for the previous one extend the construction time and cause people on site to work inefficiently. Besides, main components should have a relatively low weight to be lifted where possible by laymen and if necessary with help of simple tools, instead of big cranes, which are expensive in many rural regions, if available at all. Figure 6-7 shows an example of lifting a wall panel on site by students.

Assembly can take place in the workshop or on the site, depending to the allowable size for transportation. To provide the necessary quality of steel structures, all welds are executed in the industrial workshop, while all connections on site are bolted or screwed connections. These connections are intended further to be simple enough to eliminate the need of presence of experts in the field. All bolted connections must have enough space to be fastened by standard spanners.

6. Structural design of earthquake-safe houses and schools

Figure 6-7: Erecting wall panels on site without lifting tool

Generally, an easy-to-assemble layout can be achieved if the assembly operations are [35]:

- Structured
- Reduced
- Standardized
- Simplified

This will lead to a reduction in expenditure because the assembly process is improved and got quality-proof. In this regard automatic preparation of assembling drawings is very helpful as done by high-performance CAD systems such as Bocad-3D and Tekla Structures.

6.2.2. Corrosion-resistant design

Galvanization produces the most durable corrosion preventing system for steel structures. Galvanization means covering steel with a film of zinc by dipping it in into the melted zinc. The hot-dip galvanization process is the most appropriate (but not the only) one for structural elements and contains the steps shown schematically in figure 6-8. The first step for a good galvanization is cleaning the steel surface from oil and fats. Therefore steel elements are clipped into a caustic cleaning bath and afterwards shortly in water, to remove fats from steel. In the pickling bath, the rest of rust and scales are removed with help of diluted mineral acids. These materials are then washed off in the rinsing bath. Steel elements are then entered to a flux solution and a thin film covers the surface of elements, which eases later the reaction between steel and zinc. This film is getting dried in a drying oven and goes into the melted zinc. Although zinc melts in 419°C, operating temperature of zinc bath is usually between 440°C and 460°C or even higher. During galvanization, a layer of iron-zinc alloy covers the surface of steel as a result of an interactive diffusion between both materials. After pulling out, a layer of pure zinc as well as a layer of alloy remains on the surface of steel as shown in figure 6-9. At the end, depending to the type and thickness of the material, elements are getting cooled in the air or water.

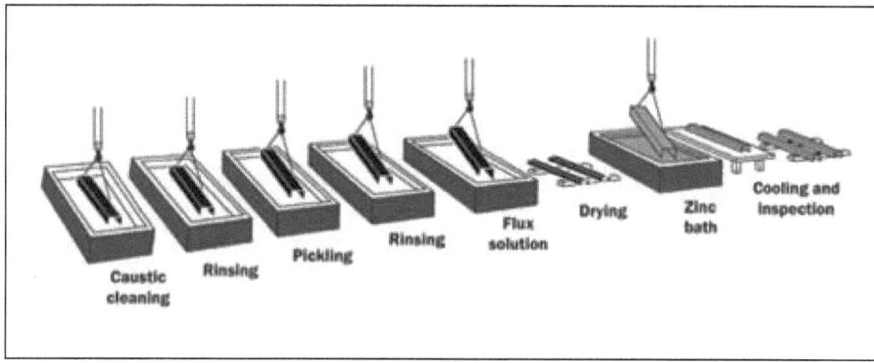

Figure 6-8: Hot-dip galvanization process [78]

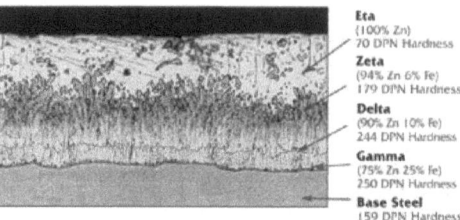

Figure 6-9: Photomicrograph of the galvanized coating [78]

The durability of the protecting layer depends to its thickness and the conditions in which the building is constructed. With help of the graph from [78] shown in figure 6-10, thickness of the protective layer can be easily concluded. For structures directly apposed to aggressive environments in the cities for a life time of 50 years, this thickness is about 80 µm. Getting galvanized in the workshop, elements will be protected not only during operation, but also in transportation and assembling.

6. Structural design of earthquake-safe houses and schools

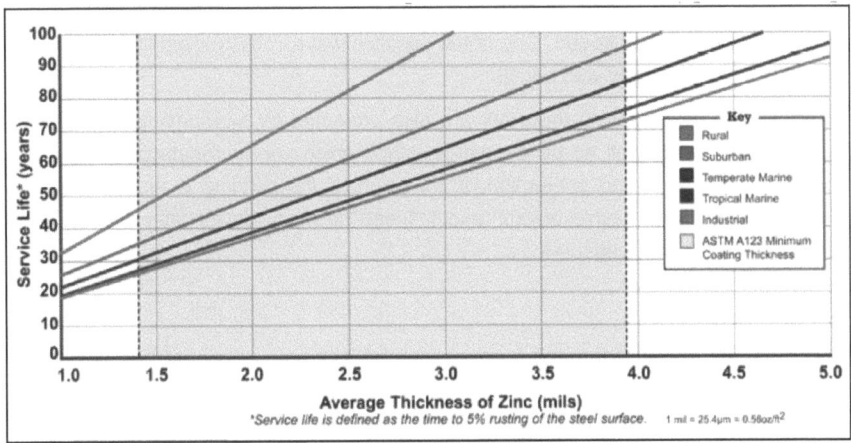

Figure 6-10: Service-life chart for hot-dip galvanized coatings [78]

Another interesting point regarding galvanization is that small damages or scratches on the surface can be repaired in a so called cathode protection by material itself, in which a barrier is built over the damaged part protecting the surface. In other words, in the place of damage, zinc is sacrificed in an electrochemical process to protect steel.

To protect steel by galvanization effectively, special care has to be paid during planning and design. Generally, assemblies like wall panels made of hot-rolled profiles are galvanized after welding. In case of cold-formed profiles, steel sheets are galvanized before forming, in order to prevent distortion of profiles. Therefore, connections have to be bolted or screwed instead of welded connections, because zinc is quite problematic for a safe welding process. The form of profiles of wall panels is also appropriate for galvanization in case of using L-profiles, since there would be no danger that the zinc remains in the profile after galvanization. If U-profiles are used, run-out holes on the corner of the panel are necessary. Below, some of the most important issues, which have to be considered for houses and schools, are presented:

- The ratio of the highest to the lowest thickness should not be more that 2.5 in an assembly and not more than 5.0 in connections.

- For welded connections, the production aspects of welding have to be considered. Welds should be clean without pores or cuts. Welding slag has to be removed in pretreatment phases of galvanization.

- Another important issue regarding welding is that the sequence of welding should be in a way to produce the least residual stresses in the structure. Therefore welds should be executed as symmetric as possible to avoid any distortion or large stress in the structure. Due to the increase of the temperature up to 450°C during galvanization, the yielding point of steel will be reduced, and larger distortions can be formed. For this reason, using automated welding machines ensure a uniform welding with a constant speed. One of the possible ways in this regard is to use welding robots for welding of wall panels.

Steel panels developed in this work are also protected by surrounding adobe or masonry materials as well as by a covering strip, which serves for decoration function, heat-flow insulation and prevention of corrosion. If the building is aimed to be constructed in non-aggressive environments, these can provide the steel structure with additional protection measures.

6.2.3. Financing

Housing may be acquired through purchase, rental, self-help or cooperative construction and inheritance. Subsidies are provided in many developed countries to households which purchase as well as those which rent. Creation of self-sustaining finance systems to meet the need for affordable finance of the people when purchasing, building or improving their dwelling units has been an important component of national and local policies for the achievement of the goal of shelter for all. Housing finance enables households to become owners, and is provided in most countries by specialized housing finance institutions and private banks. Under stable economic conditions with low levels of inflation and interest rates, housing finance systems work well. In spite of variations observed between different countries, middle and higher income groups use most of the mortgage credits. In inflationary economies, particularly in developing countries, affordable housing credit is difficult to obtain even for middle and upper-income groups. Many developing countries have therefore set up public sector housing finance institutions, to provide households with microfinance. Microfinance refers to the provision of financial services to poor or low-income clients, including consumers and the self-employed. These usually provide loans at interest rates below the market or even the inflation rate, using funds from budget allocations and captive savings in the public sector, such as reserves of insurance institutions and pension funds. The contribution of mortgage credits to housing acquisition in developing countries is much smaller than in industrialized ones, as households in the former rely predominantly on their own savings often supplemented by informal loans from friends and relatives [77].

According to the categorization of UN-HABITAT in [76], housing finance sources in developing countries generally falls into three categories. The first category is comprised of private commercial institutions providing credit for upper-income households at market interest rates, upon the certification of income streams and provision of collateral. This category of financial institutions avoid involvement in provision of housing finance for the poor due to their lack of collateral and steady income, because of high default risk, and high transaction costs.

The second category is the public sector, which usually provides subsidized funds for middle-income groups and civil servants by way of specialized or non-specialized housing finance intermediaries. In many developing countries these public housing programmes have failed to reach the poor. In developing countries, the majority of population does not qualify for mortgage finance from formal financial institutions to purchase the least expensive and economical built housing units. They are left to build their own housing units without formal financial sector support, relying on incremental financing support from non-formal financial institutions. Many developing countries even have nothing like a viable mortgage finance sector. As a result, the third group has to rely on informal sources, including savings, informal loans from friends and family, remittances from family members working abroad, and the sale of whatever assets they have.

An increasing number of institutions are becoming involved in housing microfinance services, including Microfinance Institutions (MFIs), Non-Governmental Organizations (NGOs), and Community-Based Organizations (CBOs). Table 6-4 shows the area of focus and examples of different institutions.

A vast majority of poor people cannot meet their housing needs on the open market. Therefore, even in developed countries, government plays a strong role in the housing delivery system. Government can play four types of role in the housing market: (1) An allocative role: to intervening in the allocative function of the market to improve efficiency; (2) a distributive role: as market-based distribution may not achieve equity, government may step in to align the distribution of outcomes with equity principles; (3) a stabilization role: government can stabilize the market through various steps and policies; (4) a regulatory role: the most common form of intervention in the housing sector is through regulation, direct provision and subsidies.

Due to the complex process of financing housing units in developing countries, the manufacturing enterprise can not rely on them for an extensive period of time. Instead, the enterprise should look for different institutions financing mass construction of houses. For the case of school construction, this issue looks less critical, since schools are mostly financed by governmental organizations and the budget is already allocated. The main sources of credit are not only those coming from the government, but are also provided by donors. In some countries like Iran all sources are managed by Schools Renovation, Development and Mobilization Organization, which is responsible for construction and maintenance of schools in whole country. In this case, that would be the most effective way if the organization, qualifies enterprises, which have the capacity to manufacture and construct schools in a long term following the intended standards.

Types of Institutions	Area of Focus	Examples
MFIs	Large-scale MFIs with over 100,000 clients; Housing portfolio often borne out of a disaster situation or as diversification; May be a reward for successful completion of a microenterprise loan. Medium-sized MFIs with 10,000-100,000 clients; Most have already achieved best practice in microfinance; Similar principles are applied to housing products (short term, small amounts); Some have taken government funds for expansion; Commercial funding usually not available for these loans, resulting in funding mismatch	Grameen Bank CALPIA (El Salvador, specialized finance company), BancoSol (Bolivia, bank), ADEMI (bank, Dominican Republic), MiBanco (bank, Peru), CARD Rural Bank (specialized bank, Philippines)
NGOs and CBOs	Capacity to transfer technologies across affiliates in various countries; Limited focus on technical assistance for housing products; Currently working on commercial funding for conventional micro-enterprise portfolios; Could leverage financing for housing; Some are direct lenders and some are wholesale providers of credit	Accion, CHF International, FINCA, Homeless International
Co-operatives, Mutuals and Municipals	Locally owned and often locally started housing programmes; Good experience and best practice; Usually part of networks that enable cross-experience sharing	Jesus Nazareno (S&L co-op, Bolivia), Mutual La Primera (housing co-op, Bolivia), Caja Arequipa (municipal co-op, Peru)
Government Housing Programmes	Some are professionally run; others are very political and/or not market-based; Major source of second tier financing for housing but with limited outreach; Demonstrated outreach to low income clients	Ex-FONVIS (Bolivia), FONAVIPO (El Salvador)
Commercial Banks	Some downscaling to housing faster than to microcredit; Security is a major issue; Have the capacity to expand; Could mobilize large amounts of commercial financing	Banco de Desarrollo (Chile), African Bank (South Africa)

Table 6-4: Providers of housing microfinance services [76]

6.2.4. Role of experiments

In order to verify the validity of results, two types of experiments were carried in the period of time from 2006 to 2009. These experiments are categorized and explained in the following sub-chapters.

Summer schools

During three summer schools held from 2006 to 2008 at Bergische University Wuppertal, civil engineering student from Isfahan University of Technology in Iran contributed in construction of model houses. The aim was to train students for practical work on site, mainly accomplishment of masonry and adobe walls, from one side, and to investigate the process as a whole as well as the reaction of people. These students can later bring and develop the idea of systematically braced frames in rural areas and small towns nationwide. In 2006, among other programs and visits, students were trained in two days by master masons of a construction promotion center to brick walls. Figure 6-11 shows a part of the work of students on an infill wall as well as an example of bricking pattern prepared by mason master.

 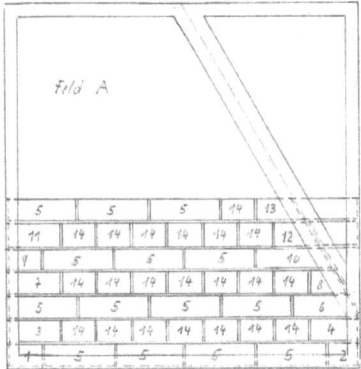

Figure 6-11: Brick work of the students after one day as well as an example of bricking pattern prepared by a mason master

The majority of trained students took part in 2007 in another summer school, where a single storey sample house of 3.5x6.0 m was erected on the parking area of Bergische University Wuppertal, based on analysis and structural design in two weeks and the infill walls of the front side were partially accomplished. Wall panels of 1.5x2.8 m were made of cold formed U-profiles of 120x60x1.5 mm. The roof was an assembly of trapezoidal sheets with maximum height of 160 mm. Figure 6-12 shows students during work and a view of the final house. The building was disassembled afterwards and wall panels were stored. A as a results of this summer school many construction related problems, including fitting of bricks into frames as well as some difficulties regarding assembling of the roof were recognized.

Figure 6-12: Students during work and a view of the final house 2008

In 2008, a two storey building of 4.5x4.5 m in plan was erected again, using a combination of wall panels from the previous year and new wall panels made of L-profiles. Some wall panels were filled with masonry bricks and some other with adobe bricks of straw loam mixture. The latter was used because it is still the most available and the least expensive material in many rural areas of developing countries. The building was also intended to be equipped with an easy friction base-isolating system. To do that, layers of galvanized steel strips were installed on a layer of polyethylene to separate all wall panels from the ground surface by decreasing the friction. The friction coefficient and sliding behavior of this system had been tested before in [74] under lower weights. However the system did not work properly with the actual weight of the building without additional decreasing of the friction by greasing. Photos in figure 6-13 illustrate the construction and a final view of the building. The building was disassembled afterwards in two days. Other highlights of this summer school were some shortcomings in the roof assembly which was constructed by timber elements for economic reasons. However, raining during construction of the building led to some problems, which demonstrated the importance of adequate isolation in the roof.

Figure 6-13: Students during construction and a final view of the building in 2009

Laboratory experiment

To investigate the behavior of a wall panel under horizontal loads, an experiment was carried out on the panel with and without infill at Bergische University Wuppertal. In this experiment the panel was tested first without infill and afterwards with infill made of straw loam bricks and mortar. The setup of experiment shown in figure 6-11 is established in a way to comply with real conditions. Due to the technical consideration including lack of a horizontal load applying system, the panel was being rotated 90 degrees and tested with a vertical load applying system. Calibrations have been made to compare and conclude results for vertical panels. The panel was assumed as a part of the first storey wall of a two storey building and vertical load of the upper storey is modeled using pre-tensioned steel bars connected to a U-profiled beam on both sides of the panel.

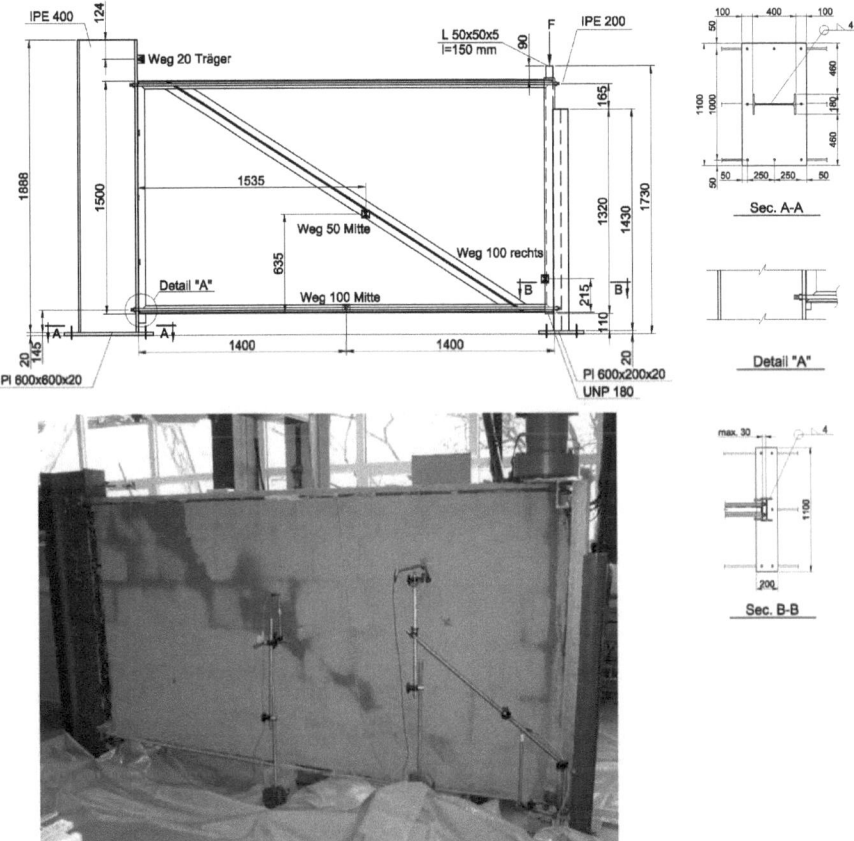

Figure 6-14: Setup of the experiment

The failure mechanism in case of frame without infill was observed as elastic buckling of columns, while in the frame with adobe infill no early instability was observed. Instead, some plastic deformations were investigated in the columns near to the connection with the beam, shown in figure 6-14. An interesting result was that no collapse happened in the adobe wall, except minor cracks in location of maximum displacements. This shows the high ductility of confined adobe walls. One important aspect, which was increasing the ductility, was moisture in the adobe at the time of experiment. Detailed information of the experiment including comparison with analytical and numerical calculations is discussed in [41].

7. Case study: A sample school in Iran

7.1. Building description

In order to show the feasibility of the developed method, a school building shown in figure 7-1 is considered and worked out. This school is designed and constructed in conventional steel structure in Isfahan, Iran.

To make the plan applicable for other regions, the seismic hazard is conservatively considered to be very high according to the seismic design code of Iran [89] with seismic ground acceleration of $a_g = 4.0 \, m/s^2$. The school consists of two stories and the first storey has a setback of 3.0 m in the front, which creates an entrance area. Above this set back there is an assembly room with a span of 7.5 m. There is also a penthouse over the staircase on the roof. Dimensions shown on the plan in figure 7-1 are modified in a way that all lengths are multiplications of 1.5 m, to comply with the developed wall panels.

All walls shown are assumed to be constructed either with braced panels or with simple ones, explained before. The steel material is taken as S235 everywhere. Panels will be connected to each other with bolts to form a complete wall. This way walls with desired lengths will be achieved with minimum effort on shipping and/or lifting tools on the site. After construction of Walls, girders and supports, i.e. T-girders, seated-T as well as seated-U profiles are assembled in a distance of 1.5 m, to prepare the support for hollow concrete plates. Different plates made of C20/25 concrete with lengths up to 4.5 m are used, to form floor and roof slabs. To make construction without temporary supports possible, the maximum span between walls has not to exceed 4.5 m. However, as evident in the figure, where the limit of span is exceeded, a beam is considered to divide the span. After construction of the first storey, the second storey will be constructed in the same manner. To support the front part of the second storey, 4 moment resisting steel frames including columns and beams are constructed on axis C to F, to show the flexibility of the system in combination with conventional systems.

Because the aim of this example is to investigate only the steel structure, panels made of L-profiles and U-profiles are compared during analysis and structural design, in sense of economic criteria.

To make the appearance of the building attractive, an architect was involved from the beginning of the case study in the project as advisor. The 3D architectural model of the school prepared as a result of this collaboration is presented in figure 7-2.

Product development of earthquake-safe houses and schools

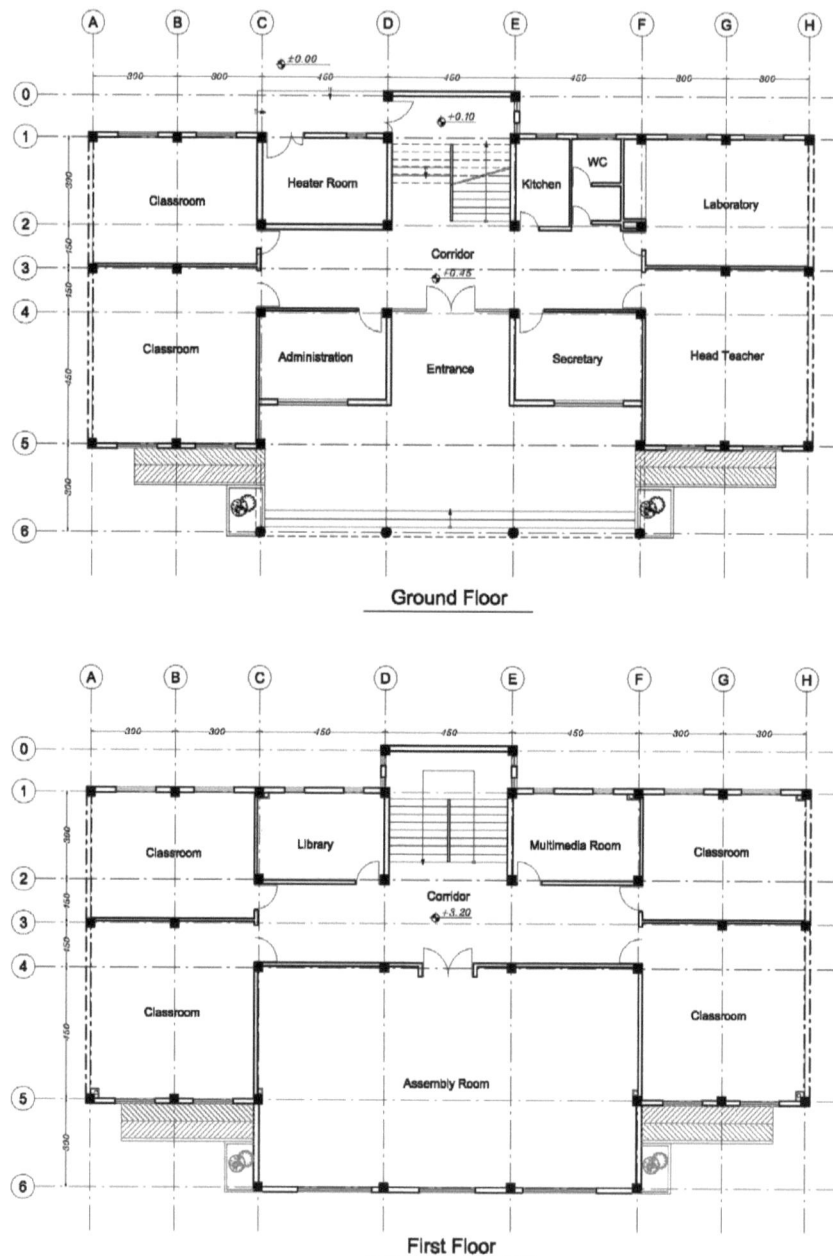

Figure 7-1: Plans of the sample school

7. Detail design

Figure 7-2: Architectural model of the sample school *[Courtesy: N. Nasserian]*

7.2. Modeling and analysis

7.2.1. Loads

Besides the self weight of the structure, extra dead loads of 1.0 kN/m² and 1.5 kN/m² are considered on the first floor and on the roof, respectively, to take the effect of floor and roof finishing into account. The live load is considered to be 3.0 kN/m² according to EN 1991-1-1 [21] for school buildings.

A bow imperfection with amplitude of $e_0/L = 1/200$ is considered according to EN 1993-1-1 [23], which implies for U-profiles. Sway imperfections are not considered due to the fact that seismic horizontal loads in each storey exceed 15% of vertical loads in that storey.

Considering the location of the school, seismic loads are assumed as elastic design spectrum type 1 for subsoil class C according to EN 1998-1 [27]. This soil class has similar characteristics to the subsoil class III in Iranian seismic design code. This spectrum (figure 7-3) can be represented by the following relations:

$$0 \leq T \leq T_B \quad S_d(T) = a_g.S.\left[\frac{2}{3} + \frac{T}{T_B}\left(\frac{2.5}{q} - \frac{2}{3}\right)\right]$$

$$T_B \leq T \leq T_C \quad S_d(T) = a_g.S.\frac{2.5}{q}$$

$$T_C \leq T \leq T_D \quad S_d(T) = a_g.S.\frac{2.5}{q}\left[\frac{T_C}{T}\right]$$

$$T_D \leq T \quad S_d(T) = a_g.S.\frac{2.5}{q}\left[\frac{T_C.T_D}{T^2}\right]$$

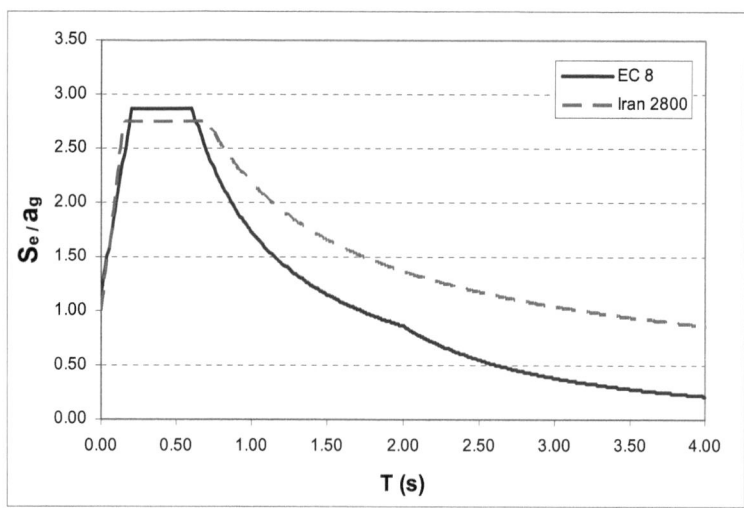

Figure 7-3: Elastic spectrum type 2 for soil class C in EN 1998-1 and soil class III in Iranian seismic code

As required in the code, seismic load is considered to influence separately in positive and negative directions along X and Y axis of the building. When considering seismic load in each direction, 30% of the seismic load in orthogonal direction is added to the load, under load case SPEC1. An additional effect using square root of sum of squares (SRSS) of the actions in both directions is considered as well, under load case SPEC 2.

After implementation of all load cases on the structure, following load combinations are implemented for analysis in the program, in which D stands for dead load, SD for extra dead load, L for live load, and I_x and I_y for bow imperfections along X and Y axis, respectively:

$Comb1 = 1.35.D + 1.35.SD + 1.5.L$

$Comb2 = D + SD + 0.3.L + Spec1$

$Comb3 = D + SD + 0.3.L - Spec1$

$Comb4 = D + SD + 0.3.L + Spec2$

$Comb5 = D + SD + 0.3.L - Spec2$

$Comb6 = 1.35.D + 1.35.SD + 1.5.L + I_x$

$Comb7 = 1.35.D + 1.35.SD + 1.5.L + I_y$

$Comb8 = D + SD + 0.3.L + Spec1 + I_x$

$Comb9 = D + SD + 0.3.L + Spec1 + I_y$

$Comb10 = D + SD + 0.3.L - Spec1 + I_x$

$Comb11 = D + SD + 0.3.L - Spec1 + I_y$

$Comb12 = D + SD + 0.3.L + Spec2 + I_x$

$Comb13 = D + SD + 0.3.L + Spec2 + I_y$

$Comb14 = D + SD + 0.3.L - Spec2 + I_x$

$Comb14 = D + SD + 0.3.L - Spec2 + I_y$

7.2.2. Modeling

The steel structure of the school is modeled in finite element program SAP2000, V.12, using beam elements. Connection of wall panel to the concrete slab is modeled as hinge connections, which can transfer shear forces but no moments, due to the fact that wall panels are connected to the slabs with only one row of bolts. For modeling of cross sections, different combinations of profiles are required at different positions as shown in figure 7-4. Geometry of the calculation model in SAP2000 is presented in figure 7-5.

7.2.3. Analysis

Doing an elastic stress analysis, the maximum forces in different sections are extracted. In tables 7-1 and 7-2 maximum forces are listed for two cases: 1) Model using only L-profiles, and 2) Model using back-to-back U-profiles for columns and bracings and L-profiles for horizontal elements. Maximum actions are resulted from different combinations and in each set the main criterion is printed in bold.

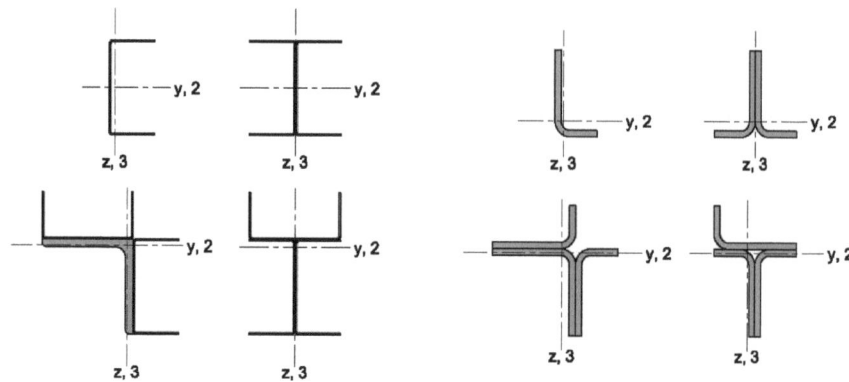

Figure 7-4:   Different frame sections created by combination of U and L-profiles

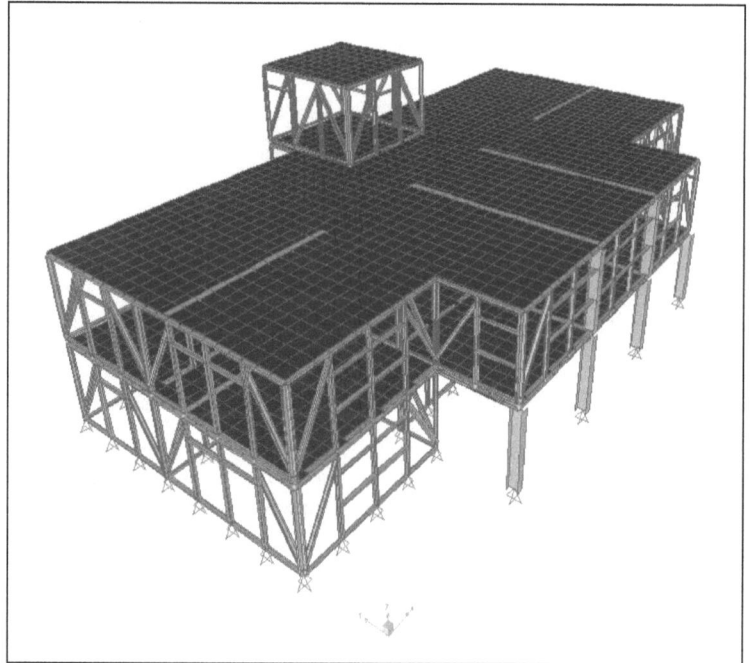

Figure 7-5:   Model of the school building in SAP2000

7. Detail design

Profile	Length [cm]	P [kN]	V2 [kN]	V3 [kN]	T [kN-cm]	M2 [kN-cm]	M3 [kN-cm]	Resistance [kN] - [kN-cm]
U120x60x2	150	0.00	0.00	0.15	0.00	19.34	0.00	85.24
		0.00	0.00	-0.17	0.00	-18.12	0.00	85.24
2xU120x60x2 (Columns)	300	6.76	0.10	4.89	0.00	94.89	6.16	163.84
		-10.01	-0.04	-3.12	0.00	-27.46	0.00	196.60
		0.37	0.09	9.93	0.00	178.62	8.19	204.53
		-2.72	-0.13	-9.14	0.00	-191.63	-6.70	204.53
		4.56	6.01	0.09	0.00	1.96	112.27	818.43
		-5.52	-5.55	-0.09	0.00	-2.00	-117.73	818.43
2xU120x60x2 (Bracings)	335	12.44	0.00	0.00	0.00	0.00	0.00	202.75
		-13.26	0.00	0.00	0.00	0.00	0.00	202.75
3xU120x60x2	300	9.68	4.14	2.32	0.00	26.96	81.49	249.03
		-13.36	-4.66	-2.29	0.00	-16.63	-7.01	249.03
		0.54	0.58	9.05	0.00	187.98	28.32	615.13
		-3.02	-0.50	-9.36	0.00	-183.37	-34.45	615.13
		1.31	0.39	0.01	0.00	9.16	86.42	961.42
		-10.17	-4.19	-0.01	0.00	-8.73	-84.00	961.42
2xU120x60x2+L 120x8	300	13.72	2.82	5.87	0.01	104.51	46.00	614.89
		-18.44	-2.75	-5.32	-0.01	-7.04	-4.23	614.89
		1.14	0.25	0.51	0.01	104.80	50.89	821.20
		-18.40	-2.75	-5.33	-0.01	-113.61	-48.01	821.20
		11.91	5.10	2.75	0.01	48.22	109.50	821.20
		-3.31	-0.48	-0.40	-0.01	-87.84	-108.71	821.20

Table 7-1: Results of analysis and capacity calculation for U-profiles

Profile	Length [cm]	P [kN]	V2 [kN]	V3 [kN]	T [kN-cm]	M2 [kN-cm]	M3 [kN-cm]	Resistance [kN] - [kN-cm]
L110x55x8	150	0.00	0.00	0.25	0.10	39.00	0.01	129.75
		0.00	0.00	-0.33	-0.09	-36.32	0.00	129.75
2xL110x55x9 (Columns)	300	16.72	0.08	0.79	0.00	33.95	7.61	209.27
		-21.35	-0.08	-6.45	0.00	-63.98	-0.22	209.27
		-0.24	0.34	19.88	0.00	357.52	7.55	387.25
		-3.04	-0.37	-18.91	0.00	-373.77	-5.96	387.25
		2.42	10.19	0.29	0.00	7.26	182.27	1162.66
		-4.90	-9.42	-0.29	0.00	-7.25	-191.59	1162.66
2xL110x55x8 (Bracings)	335	23.97	0.00	0.00	0.00	0.00	0.00	146.56
		-24.96	0.00	0.00	0.00	0.00	0.00	146.56
3xL110x55x8	300	19.75	8.00	5.26	0.02	60.38	154.17	515.94
		-29.80	-7.66	-6.79	-0.02	-38.63	-15.71	515.94
		-0.02	1.25	18.11	0.02	360.96	66.13	884.69
		-4.91	-1.17	-18.38	-0.02	-356.61	-72.88	884.69
		4.03	0.62	0.08	0.02	62.45	158.87	1302.79
		-19.54	-8.22	0.00	-0.02	-18.91	-152.51	1302.79
2xL110x55x8+L 110x8	300	25.72	6.05	9.96	0.02	168.91	106.72	874.18
		-35.33	-6.30	-9.08	-0.02	-17.83	-14.73	874.18
		-1.51	0.74	0.92	0.02	186.65	148.95	1366.48
		-32.73	-8.40	-9.93	-0.02	-188.55	-142.36	1366.48
		-1.49	1.03	0.70	0.02	146.35	203.73	1300.48
		-32.42	-11.02	-7.84	-0.02	-148.89	-202.00	1366.48

Table 7-2: Results of analysis and capacity calculation for L-profiles

To calculate the load-bearing capacity of class 4 sections used, the elastic buckling analysis program CUFSM (v. 3.12) developed by Schafer and Ádány in [79] is used, which takes local, distortional and global buckling effects into account. Results for each cross-section and corresponding loading are listed in the last column of tables 7-1 and 7-2, with consideration of interaction between axial force and moment.

As can be observed in these tables, design with U-profiles lead to more economic results and smaller sections. Here, the effect of masonry walls on the load bearing capacity of the building was excluded.

7.3. Cost calculation

A cost calculation for the main components of the steel structure with U-profiles as columns and bracings and L-profiles as horizontal elements, and for hot-rolled sections used is presented in tables 7-3 and 7-4, respectively. Prices are according to the Iranian "Fees Structure for Buildings and Structures" used for public contracts in Iran in Rials, for the Iranian calendar year from March 2008 to March 2009 [82]. Due to the rapid changes between the values of Iranian Rial and Euro, prices are not presented in Euro currency here. Instead, costs are presented in addition as relative costs explained in chapter 5, in the last columns of the tables. This value shows the proportion of the costs in comparison to the price of steel I-profiles at the same period of time.

Type of component	Number	Cost [Rials]	Total cost [Rials]	Relative cost
Braced panels	81	1,777,758	143,998,398	15,398.5
Door panels	49	1,261,587	61,817,763	6,610.5
Window panels	34	1,419,840	48,274,560	5,162.2
T-girders of 1,5 m length	16	529,222	8,467,552	905.5
T-girders of 3,0 m length	30	1,058,337	31,750,110	3,395.2
T-girders of 4,5 m length	65	1,842,647	119,772,055	12,807.8
Hollow concrete plate of 20 cm thickness and 1,5 m length	16	480,969	7,695,504	822.9
Hollow concrete plate of 20 cm thickness and 3,0 m length	30	961,938	28,858,140	3,085.9
Hollow concrete plate of 20 cm thickness and 4,5 m length	65	1,442,907	93,788,955	10,029.3
		Σ	544,423,037	58,217.8

Table 7-3: Cost calculation for the main structural components of the sample school

7.4. Highlights

The following results obtained during this case study:

1. Wall panels can be combined where necessary with hot-rolled I-sections of columns to make the execution of spans larger than 4.5 m possible. This may be required in

special cases like here for construction of schools, but for economic reasons, larger spans are better to be avoided.

Type of element	length [m]	weight [kg/m]	Total weight [kg]	Unit price [Rials]	Total price [Rials]	Relative cost
HEA 260 (Beam)	54	68.14	3679.56	10,700	39,371,292	4,210.2
HEA 260 (Column)	12.4	68.14	844.94	11,900	10,054,738	1,075.2
HEA 300 (Beam)	30	87.92	2637.6	10,700	28,222,320	3,018.0
HEA 300 (Column)	49.6	87.92	4360.83	11,900	51,893,901	5,549.3
			Σ		129,542,251	13,852.6

Table 7-4: Cost calculation for hot-rolled sections used in the sample school

2. A high effort is required for preparation of details when designing buildings with modular components for the first time. The main details were shown before in chapter 5, and were also used here. However, the required effort will be reduced as the number of designs increases. To realize all possible barriers for detailing with the new system, more buildings with different architectural layouts have to be considered. Details can be gathered as detailing guidelines for future.

3. A comparison between the original model and the model with systematically braced frames was carried out during this study. The total weight of steel required only for the main steel structure shows a reduction of about 25%. Because some modifications were implemented on the architectural layout of the school to make use of modular components possible, the structure does not completely comply with the original building. These modifications include changing the distance between some axes, which leads sometimes to smaller spans. Therefore, an exact comparison was not possible, but the high reduction in steel amount was evident.

4. In addition to the weight, the number of different steel profiles required in the design is reduced from 12 types in original design to 6 types in the modified one. Considering the large fluctuations of material, especially in developing countries, this issue is decisively important.

8. Concluding remarks and future works

While according to the statistics, the demand on new houses and schools is huge, seismic hazard and risk studies reveal that all construction activities in areas under consideration have to strictly comply with seismic design provisions. Current construction techniques are been unable to fulfill this demand and to minimize seismic risk efficiently. The high level of risk is recognized to be mainly due to corruption and/or unintentional mistakes during construction; both have to be reduced by help of technical approaches.

While many of the structural mistakes are observed to be stereotypes during previous events, some simple structural types show a surprisingly good performance. Fitting examples of such a good seismic behavior are half-timber structures, which have withstood moderate to large shakings in many regions of the world. However, the main reason for success of this type of structures depends not only to use of a ductile structural material, i.e. wood. Other contributing aspects include a high number of bracings, which increases structural ductility and divides infill walls into small fields, a relatively light weight of floors and roofs, visibility of main structural elements, which makes them provable, ease of accomplishing the non-safety-relevant works, and their affordability. As shown here, all these parameters can be used further in a systematic approach, where main elements are manufactured in a modularized industrial way with no error.

As discussed during this work, to end up with a robust solution to a real-life problem is a challenging task demanding the collaboration of several disciplines of civil engineering. Therefore, to bring all attempts into effect, a framework is required, which makes efficient collaboration of interdisciplinary solutions at different times and in different dimensions possible. This framework has also the function of implementing the research into practice and taking the last steps to bring products, i.e. houses and schools to the markets in need.

Therefore, the main four phases of the classical Product Development can be adopted for civil engineering purposes. First, in a task-clarification phase, main requirements of the product as a whole and its components have to be prioritized. During the second or conceptual design phase, the main concept of low-rise steel buildings as a promising solution is characterized. These products are suitable for small towns and rural areas in developing countries even with a high level of seismic hazard.

Furthermore, the classification between safety-relevant structural elements manufactured in the workshop, and non-structural elements, e.g. finishing works carried out by local hand workers, is emphasized. While main variants of structural elements including walls, floor/roof systems, and base isolation have to be kept to a minimum number for economical reasons, the non-structural elements can vary in different regions to bring cultural acceptance and attractiveness and to make use of local recourses including labor and materials.

Consequently, wall panels are designed from steel sections which can be easily procured in developing countries. Demand on different hot-rolled steel sections which depends to the market fluctuation is reduced by substitution of cold-formed sections. In other words, the supplying enterprise can provide and store steel sheets with a limited number of thicknesses with appropriate prices and use the material for different purposes in the long run, by forming them into different shapes. Cement required for concrete slabs, is also produced in many developing countries and is available currently in most of the regions.

The product is then designed during the third phase in accordance with required rules and provisions. Here, it is usually essential to modify existing theories in order to adopt them for new product variants. This was presented in the present work for case of lateral stiffness of wall panels and cross-sectional ductility of members. The success of these theories should be verified in future by help of numerical tools, e.g. by Finite Element simulation as well as by experiments.

In the fourth phase, a detailed design prepares layouts for production and construction. Considering the key role of high-performance CAD/CAM tools in this phase, their importance in product development should not be underestimated. CAD/CAM tools help to document the whole process, which makes necessary modifications in future easy. In addition, they increase efficiency by providing a platform for collaboration of all contributors from early stages.

The next steps for implementation of earthquake-safe houses and schools are particularly related to setting up of production lines. Here, the main barrier is the availability of equipments, for instance CNC-machines, which include a large portion of establishing costs. In this regard, authorities should support transfer of key technology features from one side and make policies for diffusion of these features to most benefit from them in long run. With help of high technology production and construction, many existing qualitative and quantitative obstacles of housing market in developing countries could be eliminated.

To prepare earthquake-safe houses and schools for the market, other important aspects have to be worked through. One of the most important aspects is the building physics, i.e., thermal and acoustic insulations, which cares for convenience of residents of a building and reduces the maintenance costs. Furthermore, both schools and houses have to be protected against fire, and corresponding details have to be prepared or modified. When all details of a house are defined, a more precise cost analysis can be carried out to compare the final price with that expected by the market. Sample houses should be also built and monitored in the long run to see which aspects have to be modified.

Last but not least, socio-political commitment and motivation of a society living in hazard are strongly required to guarantee successful implementation of results for a sustainable development instead of increasing the vulnerability for instance by migration to earthquake-prone mega cities.

References

[1] U.S. Geological Survey: Earthquakes with 1,000 or More Deaths since 1900, URL: <http://earthquake.usgs.gov/regional/world/world_deaths.php>, Accessed on 15.12.2008.

[2] Dr Mahathir bin Mohamad: Creating the right platforms for knowledge based businesses in developing countries, Speech at Isfahan University of Technology on 18.09.2006, Isfahan, Iran.

[3] Geohazards International: Toward global earthquake safety: Partnering with Geohazards International, 2004, URL: <http://www.geohaz.org/contents/publications/Geohazards%20Brochure.pdf>, Accessed on 15.12.2008.

[4] Comartin, C.; Brzev, S.; Naeim, F.; Greene, M.; Blondet, M.; Cherry, Sh.; D'Ayala, D.; Farsi, M.; Jain, S. K.; Pantelic, J.; Samant, L.; Sassu, M.: A Challenges to Earthquake Engineering Professionals. In: *Earthquake Spectra*, Vol. 20, No. 4, pp. 1049-1056, 2004.

[5] Ghafory-Ashtiany, M.: Global Blueprints for Change: Case Study of Iran's Earthquake Risk Reduction Experience, 1st edition, 2001, URL: <http://www.gadr.giees.uncc.edu/downblueprint.cfm>.

[6] Kahn, M. E.: The death toll from natural disasters: the role of income, geography, and institutions. In: *The Review of Economics and Statistics*, Vol. 87, No. 2, pp. 271–284, May 2005

[7] Parhizkar, T: Instandsetzung von Stahlbetonbauwerken im Iran, Der Einfluss der Eigenschaften von "Iran-Beton" in einem feuchtheißen Klima auf die Beständigkeit des Verbundes zwischen einem Instandsetzungssystem und dem instand zu setzenden Altbeton, Dissertation, Technischen Universität Berlin, Berlin, 2000.

[8] Bachmann, H.: Seismic Conceptual Design of Buildings: Basic principles for engineers, architects, building owners, and authorities, Biel 2002, 81 p.

[9] Earthquake Engineering Research Institute (EERI); International Association of Earthquake Engineering (IAEE): World Housing Encyclopedia (WHE): an Encyclopedia of Housing Construction in Seismically Active Areas of the World, URL: <http://www.world-housing.net>

[10] Langenbach, R.; Mosalam, K. M.; Akarsu, S.; Dusi, A.: Armature Crosswalls: A Proposed Methodology to Improve the Seismic Performance of Non-Ductile

Reinforced Concrete Infill Frame Structures, In: *8th U.S. National Conference on Earthquake Engineering (8NCEE), San Francisco 1906 anniversary*, 2006.

[11] EERI Learning from Earthquakes Reconnaissance Report: The Boumerdes, Algeria, Earthquake of May 21, 2003. Earthquake Engineering Research Institute (EERI), October 2003, URL: <http://www.eeri.org/site/images/lfe/pdf/algeria_20030521.pdf>, Accessed on 15.12.2008.

[12] Turkey-US geotechnical earthquake engineering reconnaissance team: Initial Geotechnical Observations of the August 17, 1999, Kocaeli Earthquake, September 1999, URL: <http://research.eerc.berkeley.edu/projects/GEER/GEER_Post EQ Reports/Kocaeli_1999/Report3.1Kocaeli.pdf>, Accessed on 15.12.2008.

[13] Taiwan's National Center for Research on Earthquake Engineering (NCREE): Sichuan Earthquake 2008 Reconnaissance Photo Gallery V2, URL: <http://w3.ncree.org/ZH/EarthquakeInfo/080512Sichuan/D/11/index.html>, Accessed on 15.12.2008.

[14] Eslimy-Isfahany, S. H. R.; Pegels, G.: Seismic-Proof Housing: Reconstruction in Rural Areas. In: *Third International I-Rec Conference, Post-disaster reconstruction: Meeting Stakeholder Interests,* 17-19 May 2006, Florence, Italy, pp. 335-341.

[15] Ledezma, Ch.: Preliminary overview report of M 7.7 Antofagasta-Tocopilla, Chile Earthquake of November 14, 2007, UC Berkeley. URL: <http://www.eeri.org/lfe/pdf/chile_nov07_ledezma.pdf>, Accessed on 15.12.2008.

[16] National Geophysical Data Center (NGDC): NGDC Natural Hazards Slide Sets, URL: <http://www.ngdc.noaa.gov/nndc/struts/results?eq_0=5&t=101634&s=0&d=1>, RAccessed on 15.12.2008.

[17] Goldbeck GmbH: Construction business abroad, opportunities for medium-sized enterprises (In German: Auslandsbau - Chancen für den Mittelstand), Presentation, Bergische Universität Wuppertal, Wuppertal, December 6th, 2007.

[18] Steel Framing Alliance, URL: http://www.steelframing.org/, Accessed on: 15.12.2009.

[19] Samsamshariat, M.: Vulnerability of Predominant Building Types in Iran, Master Thesis, Faculty of Civil Engineering, Bauhaus-University Weimar, Weimar, 2005.

[20] European Committee for Standardization: prEN 1990: 2001: Basis of structural design, Brussels, 2003.

References

[21]   European Committee for Standardization: EN 1991: 2002: Actions on structures; Part 1-1: General actions: Densities, self-weight, imposed loads for buildings; German version, Beuth Verlag, Berlin, 2002.

[22]   European Committee for Standardization: EN 1992-1: 2002: Design of concrete structures, Part 1: General rules and rules for buildings, Brussels, 2002.

[23]   European Committee for Standardization: EN 1993: 2003: Design of steel structures; Part 1-1: General rules and rules for buildings, Brussels, 2003.

[24]   European Committee for Standardization: EN 1993: 2004: Design of steel structures; Part 1-3: Supplementary rules for cold-formed members and sheeting, Brussels, 2004.

[25]   European Committee for Standardization: EN 1993: 2004: Design of steel structures; Part 1-5: Plated structural elements, Brussels, 2004.

[26]   European Committee for Standardization: EN 1993: 2003: Design of steel structures; Part 1-8: Design of joints, Brussels, 2003.

[27]   European Committee for Standardization: EN 1998-1: 2003: Design of structures for earthquake resistance, Part 1: General rules, seismic actions and rules for buildings, Brussels, 2003.

[28]   European Committee for Standardization; EN 1337-3:2005: Structural bearings; Part 3: Elastomeric bearings; German version, Beuth Verlag, Berlin, 2005.

[29]   Nakata, J. K.; Meyer, Ch. E.; Wilshire, H. G.; Tinsley, J. C.; Updegrove, W. S.; Peterson, D. M.; Ellen, S. D.; Haugerud, R. A.; McLaughlin, R. J.; Fisher, G. R.; Diggles, M. F.; The October 17, 1989, Loma Prieta, California, Earthquake— Selected Photographs. In: *U.S. Geological Survey: Digital Data Series DDS-29 Version 1.2*,1999, URL: <http://pubs.usgs.gov/dds/dds-29>, Accessed on 15.12.2008.

[30]   Kelly, J. M.: Seismic Isolation Systems for Developing Countries. In: *Earthquake Spectra*, Vol. 18, Issue 3, pp. 385-406, August 2002.

[31]   Building and Housing Research Center of Iran (BHRC), URL: <www.bhrc.ac.ir>, Accessed on 15.12.2008.

[32]   Swain, Th. M.; Schwarz, J.; Burkhardt, A.; Friedrich, T.; Werner, F: Project Steel Earthquake Design (SEQD) - Erdbebenresistentes Stahlbausystem für Wohnhäuser. In: Bautechnik, Volume 82, Issue 8, pp. 549 – 558, August 2005.

[33]   International Institute of Earthquake Engineering and Seismology (IIEES), URL: <http://www.iiees.ac.ir/English/index_e.asp>.

[34] Cedillos, V.: Photo Gallery of Wenchuan, China Earthquake of May 12, 2008. URL: <http://picasaweb.google.com/vcedillos>, Accessed on 15.12.2008

[35] Pahl, G.; Beitz, W.; Feldhusen, J.; Grote, K. H.; (Translators from German and Editors: Wallace, K.; Blessing, L.): Engineering Design: A Systematic Approach, 3rd edition, Springer-Verlag, London, 2007, ISBN 978-1-84628-318-5.

[36] Naeim, F. (Editor): The Seismic Design Handbook, Springer, 2nd edition, 2001, ISBN 0-7923-7301-4.

[37] Stahl-Informations-Zentrum (Editor): Document 560: Houses in lightweight steel structures (In German: Häuser in Stahl-Leichtbauweise), 1st Edition, Düsseldorf, 2002, ISSN 0175-2006.

[38] Langenbach, R.: "Crosswalls" instead of Shearwalls: A Proposed Research Project for the Retrofit of Vulnerable Reinforced Concrete Buildings in Earthquake Areas based on Traditional *Hımış* Construction, *Fifth National Conference on Earthquake Engineering,* 26-30 May 2003, Istanbul, Turkey, Paper No: AE-123.

[39] Imhof, M.; Historical half-timber structures, (In German: Historisches Fachwerk), Dissertation, Otto-Friedrich Universität Bamberg, Bamberg, 1996, ISBN 3-87052-796-X.

[40] Hebel building system for economic construction, URL: <http://www.hebel.de/images/deu/rd_de/Wibsys-2005.jpg>, Accessed on 15.12.2009.

[41] Peyvandi, P.: Residential buildings with steel braced structures: Behavior of infill walls during and after earthquake, (In German: Wohnbauten in Stahlfachwerkbauweise: Verhalten von massiven Ausfachungen während und nach Erdbeben), Dissertation, Bergische Universität Wuppertal, Wuppertal.

[42] Gioncu, V.; Mazzolani, F. M.: Ductility of Seismic-Resistant Steel Structures, Taylor & Francis Group, London, 2002, ISBN 0-419-22550-1.

[43] Könke, C.; Crumple zone effect in steel structures provides earthquake safety; (In German: Knautschenzoneneffekt sorgt bei Stahlbauten für Erdbebensicherheit), VDI Nachrichten, Nr. 39, pp. 18-19, 26.09.2008.

[44] Website of Xella-Group, URL: <http://www.xella-group.com/html/com/en/index.php>, Accessed on 15.12.2009

[45] Goldbeck magazine: High-quality concrete for Europe (In German: Hochleistungsbeton für Europa), Nr. 38, pp. 22-23, Autumn 2008.

[46] Ytong Porenbeton: Technische Daten, URL: <http://www.hebel.de/downloads/deu/broschures/Ytong_Technische_Daten_2009.

pdf>, Accessed on 15.12.2008.

[47] Popov, E. P.; Kasai, K.; Engelhardt, M.D.: Advances in design of Eccentrically Braced Frames. In: *Earthquake Spectra*, Vol. 3, No. 1, 1987, pp. 43-55.

[48] Engelhardt, M. D.; Popov, E. P.: On Design of Eccentrically Braced Frames. In: *Earthquake Spectra*, Vol. 5, No. 3, pp. 495-511, 1989.

[49] Echo hollow concrete floors. URL: <http://www.echofloors.co.za/>.

[50] Normenausschuss Bauwesen (NABau) im DIN: DIN 1045-1: 2008, Concrete, reinforced and prestressed concrete structures. Part 1: Design and construction, German Version, Beuth Verlag, Berlin, 2008.

[51] Bindseil, P.: Reinforced concrete precast elements: Design – Calculation - Construction, (In Germany: Stahlbetonfertigteile: Konstruktion – Berechnung – Ausführung), 3rd Edition, Werner Verlag, 2007, ISBN 978-3-8041-4463-7.

[52] Köco Stud Welding; URL: <http://www.bolzenschweisstechnik.de/english/index.php?seite=index>, Accessed on 20.02.2009.

[53] Nord-Lock maximum security for bolted joints; URL: <http://www.nordlock.com/>, Accessed on 20.02.2009.

[54] Jamali, N.: On the Numerical Simulation of Friction-Isolated Structures, Dissertation, Bergische Universität Wuppertal, Wuppertal, 2008.

[55] Azadnia, H.; Transfer of high-level CAD/CAM technology to developing countries, Dissertation, Bergische Universität Wuppertal, Wuppertal, 2008.

[56] United Nations Centre for Human Settlements (Habitat); 2001; THE STATE OF THE WORLD'S CITIES REPORT 2001, ISBN 92-1-131476-3, Nairobi, Kenya, http://ww2.unhabitat.org/Istanbul+5/statereport.htm, Accessed on 20.02.2009.

[57] The Executive Director of UN-HABITAT, Mrs. Anna Tibaijuka, URL: <http://www.unhabitat.org/content.asp?cid=5809&catid=237&typeid=6&subMenuId=0, http://www.un.org/ga/Istanbul+5/30.pdf>, Accessed on 20.02.2009.

[58] United Nations, The Millennium Development Goals Report 2008: 2008, New York, ISBN 978-92-1-101173-9, URL: <http://www.un.org/millenniumgoals/2008highlevel/pdf/newsroom/mdg reports/MDG_Report_2008_ENGLISH.pdf>, Accessed on 20.02.2009.

[59] Ben Wisner; Let Our Children Teach Us!; A Review of the Role of Education and Knowledge in Disaster Risk Reduction; Final Report: 1 May 2006; URL:

[60] BBC News: Quake 'claimed 17,000 children'; URL: <http://news.bbc.co.uk/2/hi/south_asia/4393584.stm>, 31 Oct. 2005, Accessed on 20.02.2009.

<http://www.interragate.info/cogss>; Accessed on 20.02.2009.

[61] United Nations Development Programme; 2007; Human Development Report 2007/2008; Fighting climate change: Human solidarity in a divided world; ISBN 978-0-230-54704-9, Palgrave Macmillan, New York; URL: <http://hdr.undp.org/en/media/HDR_20072008_EN_Complete.pdf>; Accessed on 20.02.2009.

[62] Emergency Events Database (EM-DAT): The OFDA/CRED International Disaster Database, URL: <www.emdat.be>; Université Catholique de Louvain; Brussels; Belgium.

[63] The world Bank: 2007: Gross national income per capita 2007, Atlas method and PPP, URL: <http://siteresources.worldbank.org/DATASTATISTICS/Resources/GNIPC.pdf>; Accessed on 20.02.2009.

[64] The World Bank; Global Purchasing Power Parities and Real Expenditures; 2005 International Comparison Program; Methodological Handbook; URL: <http://web.worldbank.org/>, Accessed on 20.02.2009.

[65] Ghafory-Ashtiany; M.; 2005; Earthquake Risk Management Strategies: The Iranian Experience; URL: <http://www.iiees.ac.ir/English/Disaster/Khatarpaziri.pdf>, Accessed on 20.02.2009.

[66] Giardini; D.; Grünthal; G.; Shedlock; K.; Zhang; P.; 1999; Global Seismic Hazard Assessment Program (GSHAP): Global Seismic Hazard Map, URL: <http://www.seismo.ethz.ch/GSHAP/global/>, Accessed on 20.02.2009.

[67] Grünthal, G., 1998, European Macroseismic Scale 1998 (EMS-98), European Seismological Commission, Luxembourg.

[68] International Strategy for Disaster Reduction; Disaster statistics 1991-2005; URL: <http://www.unisdr.org/disaster-statistics/impact-killed.htm; Accessed on 15.12.2008.

[69] The New York Times of November 21, 2008; Garbled Report on Sichuan Death Toll Revives Pain; URL: <http://www.nytimes.com/2008/11/22/world/asia/22quake.html?hp>; Accessed on 20.02.2009.

[70] Naeim; F.; Brzev; S.; Advanced technologies in housing construction; World

Housing Encyclopedia; URL: <http://www.world-housing.net/>; Accessed on 20.02.2009.

[71] Schafer, B.W.; Ádány, S.: Understanding and classifying local, distortional and global buckling in open thin-walled members, *Proceedings of the Structural Stability Research Council Annual Stability Conference*, May 2005. Montreal, Quebec, Canada, pp. 27-46.

[72] Naeim, F.; Kelly, J. M.: Design of seismic isolated structures: from theory to practice, John Wiley & Sons, New York, 1999, ISBN 0-471-14921-7.

[73] Jamali, N.: On the numerical simulation of friction-isolated structures, Instituts für Konstruktiven Ingenieurbau, 2008, ISBN 978-3-940795-11-3.

[74] Rezaido, K.: Safety slip clutch for dwellings in earthquake-prone areas, (In German: Sicherheits-Rutschkupplung für erdbebengefährdete Wohnhäuser), Diploma thesis, Bergische Universität Wuppertal, Wuppertal, 2008.

[75] Maurer Söhne: MAURER Earthquake protection, (In German: MAURER Erdbebenschutz); URL: <http://www.maurer-soehne.de>, last Accessed on 04.03.2009.

[76] United Nations Human Settlement Programme (UN-HABITAT); Housing for All: The Challenges of Affordability, Accessibility and Sustainability; The Experiences and Instruments from the Developing and Developed Worlds; Nairobi; 2008.

[77] United Nations Centre for Human Settlements (Habitat): The States of the World's Cities 2001, United Nations Centre for Human Settlements (Habitat), Publications Unit, Nairobi, 2001, ISBN 92-1-131476-3.

[78] American Galvanization Association; URL: <http://www.galvanizeit.org/>, Accessed on 20.02.2009.

[79] Schafer, B.W., Ádány, S. "Buckling analysis of cold-formed steel members using CUFSM: conventional and constrained finite strip methods." Eighteenth International Specialty Conference on Cold-Formed Steel Structures, Orlando, October 2006.

[80] The International Bank for Reconstruction and Development / The World Bank and Columbia University: Natural Disaster Hotspots; *A Global Risk Analysis,* Washington, DC, 2005, ISBN 0-8213-5930-4.

[81] Blondet, M.; Villa Garcia; M.; Brzev, S.: EERI, Earthquake-Resistant Construction of Adobe Buildings: A Tutorial; Published as a contribution to the EERI/IAEE World Housing Encyclopedia, URL: <www.world-housing.net>; March 2003, Accessed on

20.02.2009.

[82]　President Deputy Strategic Planning and Control: Fee Structure for Buildings and Structures, 2008

[83]　President Deputy Strategic Planning and Control: Price list for steel, cement during the first, second and third quarters of Iranian year 1387

[84]　Jaspert, J. P.; Meyer, T.; Mononen, T.; Schneider, R.; Toma, T.; White, D.; Leino, T:Novel jointing systems for the automated production of light gauge steel elements, European Commission, 2005.

[85]　Kesti, J.; Rodriguez-Ferran, A.; Pastor, N.; Arnedo, A.; Casafont, M.; Bretones, M. A.; Arola, J.; Hakola, I.; Fulop, L.; Sivill, A.: Seismic design of light gauge steel framed buildings, European Commission, 2007.

[86]　International Conference of Building Officials: 1997 Uniform Building Code, 1997.

[87]　Johansson, B.; Maquoi, R.; Sedlacek, G.; Müller, C.; Beg, D.: Commentary and Worked Examples to EN 1993-1-5 "Plated Structural Elements", First Edition, European Commission, October 2007, ISSN 1018-5593.

[88]　Stowell, E. Z.: A unified theory of plastic buckling of columns and plates, Technical Note No. 1556, NACA, Washington, 1948.

[89]　Iranian Code of practice for Seismic resistant Design of Buildings, Standard. No.2800, 3rd edition, Building and Housing Research Center, Tehran, Iran, 2005.

[90]　Rudolstädter Systembau GmbH; URL: <www.rsb-stahlbau.de>, Accessed on 22.08.2009

Die VDM Verlagsservicegesellschaft sucht für wissenschaftliche Verlage abgeschlossene und herausragende

Dissertationen, Habilitationen, Diplomarbeiten, Master Theses, Magisterarbeiten usw.

für die kostenlose Publikation als Fachbuch.

Sie verfügen über eine Arbeit, die hohen inhaltlichen und formalen Ansprüchen genügt, und haben Interesse an einer honorarvergüteten Publikation?

Dann senden Sie bitte erste Informationen über sich und Ihre Arbeit per Email an *info@vdm-vsg.de*.

Sie erhalten kurzfristig unser Feedback!

VDM Verlagsservicegesellschaft mbH
Dudweiler Landstr. 99
D - 66123 Saarbrücken

Telefon +49 681 3720 174
Fax +49 681 3720 1749

www.vdm-vsg.de

Die VDM Verlagsservicegesellschaft mbH vertritt

Printed by Books on Demand GmbH, Norderstedt / Germany